Revis

A-level Mathematics

F G J Norton

HLT Publications

HLT PUBLICATIONS
200 Greyhound Road, London W14 9RY

First Published 1997

© FJG Norton 1996

All HLT publications enjoy copyright protection and the copyright belongs to The HLT Group Ltd.

All rights reserved. No part of this publication may be reproduced or transmitted in any form or by any means, electronic, mechanical, photocopying, recording or otherwise, or stored in any retrieval system of any nature without either the written permission of the copyright holder, application for which should be made to The HLT Group Ltd, or a licence permitting restricted copying in the United Kingdom issued by the Copyright Licensing Agency.

Any person who infringes the above in relation to this publication may be liable to criminal prosecution and civil claims for damages.

ISBN 1 0 7510 0615 7

British Library Cataloguing-in-Publication.

A CIP Catalogue record for this book is available from the British Library.

Printed and bound in Great Britain.

CONTENTS

Introduction	page	ix
The core syllabus for Mathematics A-level		xi
Addresses of Examination Boards		xiv
General examination hints		xv

1 Functions

1.1	Notes	1
1.2	Composite functions	3
1.3	Composite functions; range; inverse function	3
1.4	Range; inverse function; composite function; points of inflexion	5
1.5	Sketch; inverse function; composite function; one-to-one function	8
1.6	Sketch of modulus function; inequality	10
1.7	Deducing sketch of $y = f(x-1)$ and $y = 2f(\frac{1}{2}x)$ from $y = f(x)$	11

2 Polynomials

2.1	Notes	13
2.2	Factor theorem; factorising	13
2.3	Factor theorem; inequality	15
2.4	Remainder theorem	16
2.5	Factor theorem; solving a cubic by factorising	17

3 Quadratic equations

3.1	Notes	19
3.2	Solution of a quadratic; substitution $y = \cos x$	20
3.3	Solution of a quadratic; substitution $y = e^x$, logarithms	21
3.4	Solution of two simultaneous equations; one linear and one quadratic	21
3.5	Condition that a quadratic equation has two real distinct roots	22
3.6	Sum and product of the roots of a quadratic equation	23
3.7	Quadratic inequalities	24

4 Arithmetic series, geometric series (progressions)

4.1	Notes	26
4.2	Sum of an arithmetic progression	27
4.3	General term; sum of an arithmetic progression	27
4.4	Sum of a geometric progression	28
4.5	Sum of a geometric series; sum of an arithmetic series; sum to infinity of a geometric series	29

MATHEMATICS REVISION WORKBOOK

4.6	Sum of an arithmetic series; sum of a geometric series; sum to infinity of a geometric series	30
4.7	Condition for a geometric series to converge to a fixed value	32

5 Indices and logarithms

5.1	Notes	34
5.2	Positive and negative indices	35
5.3	Rational indices	35
5.4	Surds (square roots) and logarithms	36
5.5	Simultaneous equations involving logarithms	37
5.6	Simultaneous equations involving indices and logarithms	38

6 Summation of series

6.1	Notes	39
6.2	Summation using formulae	39
6.3	Summation using identity	40
6.4	Summation using partial fractions	42

7 Partial fractions

7.1	Notes	44
7.2	Denominator of the form $(x-a)(x-b)$	44
7.3	Denominator of the form $(x-a)(x^2+b)$	45
7.4	Denominator of the form $x^2(x+a)$; application to integrating	46
7.5	Denominator of the form $(x+a)(x+b)$; expansion in series	49

8 Binomial, logarithmic, exponential expansions

8.1	Notes	51
8.2	Binomial expansion (positive index)	52
8.3	Binomial expansion (negative index); approximate value of $1/\sqrt{3}$	53
8.4	Binomial expansion (positive integer index)	54
8.5	Binomial expansion (negative integer index); logarithmic expansion	55
8.6	Exponential expansion	55

9 Sector of a circle

9.1	Notes	58
9.2	Perimeter and area of a circle	58
9.3	Perimeter and area; maximum value of the area	59

10 Sine and cosine formulae

10.1	Notes	62

CONTENTS

10.2	Cosine formula	62
10.3	Cosine formula; deducing $\sin\theta$ from $\cos\theta$	63
10.4	Sine formula; small increment	64

11 Sums and products formulae; solution of trigonometric equations

11.1	Notes	68
11.2	Sin $(A+B)$; expansion; error	70
11.3	Equations of the form $a\cos x + b\sin x = c$	71
11.4	Equations that can be reduced to quadratics	72
11.5	Equations that can be reduced to quadratics; use of sums and products formulae	73
11.6	Equations requiring use of sums and products formulae	75
11.7	Equations requiring use of $\sin 3\theta$ formula; general solution in degrees	75
11.8	General solutions in radians	76
11.9	Use of $\sin 3\theta$ to solve a cubic	77

12 Coordinate geometry of the straight line and circle

12.1	Notes	79
12.2	Straight line through two points; length of a line-segment	80
12.3	Straight line through two points; two perpendicular lines; gradient and ratio	81
12.4	Straight line through two points; perpendicular lines; point of intersection of two lines; ratio of lengths	83
12.5	Equation of a circle; gradient of a line; area of a triangle and of a sector; gradient of a tangent	85

13 Coordinate geometry using parameters

13.1	Notes	88
13.2	The rectangular hyperbola $x = ct$, $y = c/t$	89
13.3	The curve $x = t^3$, $y = 3t^2$; finding dy/dx	90
13.4	The curve $x = t + e^t$, $y = t + e^{-t}$; finding dy/dx	91
13.5	The curve $x = t^2 - 2$, $y = t$; cartesian equation of a circle	92
13.6	Parameters in trigonometric form; finding dy/dx and d^2y/dx^2	94

14 Differentiation

14.1	Notes	97
14.2	Derivative of $\ln f(x)$; derivative of a quotient	99
14.3	Derivative of a function of a function; derivative of a product; dy/dx when y is an implicit function of x	100
14.4	Derivative of a polynomial; point of inflexion	101

MATHEMATICS REVISION WORKBOOK

14.5	Derivative of an exponential; second derivative	103
14.6	Equation of a tangent to a curve; derivative of a quotient	104

15 Integration

15.1	Notes	107
15.2	Deducing an integral from a derived function	109
15.3	Integration by inspection; by parts; by substitution	111
15.4	Use of partial fractions; inverse trig and logarithmic forms	113
15.5	Gradients and areas	115
15.6	Minimum point; area of a region	116
15.7	Area of a region: volume of solid of revolution	120

16 Differential equations

16.1	Notes	122
16.2	Integration by parts; differential equation with separable variables	124
16.3	Separable variables; inverse trig form	125
16.4	Tangent field; family of curves	125

17 Approximations to integrals

17.1	Notes	130
17.2	Trapezium rule, given $y = f(x)$	130
17.3	Simpson's rule, given the ordinates	132
17.4	Simpson's rule, given $y = f(x)$	133

18 Iterative methods of solving equations

18.1	Notes	135
18.2	Locating a root; solving using Newton–Raphson method	136
18.3	An iteration of the form $x_{r+1} = f(x_r)$	137
18.4	An iteration of the form $x_{r+1} = f(x_r)$; failure to give a specified root; an alternative iteration	138
18.5	Newton–Raphson method; comparison with $x_{r+1} = f(x_r)$	140
18.6	Newton–Raphson method; binomial expansion	142

19 Complex numbers

19.1	Notes	144
19.2	Modulus and argument of a complex number	145
19.3	Expressing a complex number in the form $a + ib$; modulus-argument form	146
19.4	Use of an Argand diagram; finding the argument of a complex number	147
19.5	Roots of a quadratic equation	150

| 19.6 | Find all three roots of a cubic equation, given one complex root | 152 |
| 19.7 | Finding a locus | 153 |

20 Vectors

20.1	Notes	155
20.2	Components of equal vectors	156
20.3	Equations of two straight lines; angle between two lines	157
20.4	Displacement vectors and the angle between two vectors	158
20.5	Vectors written in matrix form; equations of two straight lines; angle between two straight lines	160
20.6	Perpendicular lines; image of a point	162
20.7	Equation of a plane; angle between a line and a plane	164
20.8	Equation of two straight lines; position vector of a point of intersection; angle between two straight lines; distance of a point from a line	165
20.9	Equation of a straight line; finding a line through a given point perpendicular to a given line; length of a line-segment	169

21 Probability

21.1	Notes	172
21.2	Independent events. Given $P(A), P(A \cup B)$, find $P(B)$, etc	173
21.3	Dependent events. Given $P(A), P(B), P(B\,|\,A)$, find $P(B'\,|\,A)$, etc	175
21.4	Use of Venn diagrams	176

| Index | 179 |

INTRODUCTION

This book is designed to help students answer A-level Mathematics questions. It contains 100 questions, almost all taken from past A-level examination papers, together with hints, solutions and comments.

It is the intention that students should attempt the examination questions themselves, then compare their solutions and answers with those given. To achieve this, each question is printed as in the examination paper, followed by hints in the harder questions. The solutions given here have enough explanation for students to understand the method and to note the points to spot, and in almost all cases much more explanation is given than is needed in the examination. As far as the examiner is concerned, the shorter the solution the better, provided the method is made clear!

In some questions more than one method of solution is shown. In other questions, only the simplest method is given. Students should enjoy finding elegant solutions to problems which have such solutions, but these are rarely found in A-level papers.

All examination boards permit the use of calculators and students should have, and know how to use efficiently, the best available at this level. The cost, now around £50 for a good graphics calculator, is small compared with the benefits of the place at a good university that a good grade at A-level can bestow. In the section on Numerical Methods the student is given some examples how to use these efficiently; more examples will be found in the booklets accompanying the calculator, more still will be found by the students themselves with practice.

At present about 50% of all A-level syllabuses must be devoted to the common core prescribed by the Schools and Colleges Assessment Authority (SCAA). This book covers the common core and some other Pure Mathematics topics, such as complex numbers and vectors, which are found in almost all A-level syllabuses.

Looking at many dozens of recent A-level mathematics papers, one is struck by the range and variety of the questions that are set. A selection of only 100 cannot hope to cover all possible questions that can be set, but it does cover a large number of topics and the confidence it is intended that students will gain should help them to tackle new questions when they are confronted by them in the examination.

The questions used have been taken from past examination papers set by:

The Associated Examining Board (AEB)
The University of Cambridge Local Examination Syndicate (C)
The University of London Examination and Assessment Council (L)
The Mathematics in Education and Industry Project (MEI)
The Oxford and Cambridge Schools Examination Board (O & C)
The Welsh Joint Examination Council (WJEC)

MATHEMATICS REVISION WORKBOOK

I am grateful to these Boards for permission to use the questions. The solutions and answers given are entirely my own, and the Boards accept no responsibility whatever for the accuracy or method of working or the answers given.

F G J Norton

THE CORE SYLLABUS FOR MATHEMATICS A-LEVEL

The Schools and Colleges Assessment Authority (SCAA) has prescribed a core for all A-level Mathematics Examinations. This core must be included in all examinations that carry the title Mathematics, and it must comprise about half of any A-level Mathematics syllabus. The first 9 sections are usually examined in 'Pure' Mathematics papers, the last 2 partly in 'Pure' papers, partly in 'Applied' papers. The order in which topics are listed here is the order in which they occur in the SCAA listing, which is not the same as that used by examination Boards. Obviously, a thorough knowledge of these topics is vital for a good grade at A-level.

1   Algebra

Laws of indices (including negative and rational exponents)
Addition, subtraction, multiplication and factorisation of polynomials
The factor theorem
The use of the modulus sign
Solution of linear and quadratic inequalities and equations
Solution of simultaneous linear equations

2   Coordinate geometry

Cartesian coordinates in two and three dimensions
Distance between points
Equation of a straight line in the $y = mx + c$ form
Finding an equation of a linear graph
Cartesian and parametric equations of curves

3   Function

Graphical representation of functions and their inverses
Understanding of domain, range and composition
Knowledge of the effect of simple transformations on the graph of $y = f(x)$ as represented by $y = af(x), y = f(x) + a, y = f(x + a), y = f(ax)$

4   Sequences and series

Definition of sequences. Recognition of periodicity, oscillation, convergence and divergence
Arithmetic and geometric series. The sum to infinity of a convergent geometric series
Binomial expansion of $(1 + x)^n$ for positive integer n. The notations $n!$ and $\binom{n}{r}$

5   Trigonometry

Radian measure. The formulae $s = r\theta$ and $A = \frac{1}{2}r^2\theta$
Sine, cosine and tangent functions. Their graphs, symmetries and periodicity
Formulae for $\sin(A \pm B)$, $\cos(A \pm B)$. Double angle formulae

MATHEMATICS REVISION WORKBOOK

6 Exponentials and logarithms

Definition and properties of e^x and $\ln x$, including their graphs
Laws of logarithms
Exponential growth and decay
The solution of equations of the form $a^x = b$

7 Differentiation

The concept of a derivative as gradient. Differentiation and its relation to the idea of a limit
Algebraic differentiation of polynomials
Differentiation of e^x, $\ln x$, $\sin x$, $\cos x$ and $\tan x$
Application of differentiation to gradients, maxima and minima and stationary points
Increasing and decreasing functions. Rates of change
Differentiation of functions generated using sums, differences, products, quotients and composition
Differentiation of inverse functions: $\dfrac{dy}{dx} = \dfrac{1}{dx/dy}$
Formation of simple differential equations

8 Integration

The concept of integration. The evaluation of area under a curve
Integration as the inverse of differentiation
Integration of $x^n, e^x, 1/x. \sin x, \cos x$
Evaluation of definite integrals with fixed limits
Simple cases of integration by substitution and by parts
Solution of simple differential equations by either analytical or numerical means

9 Numerical methods

Absolute and relative errors in calculations when data are not known (or stored) precisely
Location of roots of $f(x) = 0$ by considering changes of sign of $f(x)$
Approximate solution of equations using iterative methods

10 Mathematics of uncertainty

Appreciation of the inherent variability of data
Collection, ordering, and presentation of data
Calculation and interpretation of appropriate summary measures of the location and dispersion of data
Combined probabilities of both independent and dependent events

11 Applications of mathematics

Understanding the process of mathematical modelling with reference to one or more application areas
Abstraction from a real-world situation to a mathematical description
The selection and use of a simple mathematical model to describe a real-world situation

THE CORE SYLLABUS FOR MATHEMATICS A-LEVEL

Approximation, simplification and solution
Interpretation and communication of mathematical results and their implications in real-world terms
Progressive refinement of mathematical models

ADDRESSES OF EXAMINATION BOARDS

The addresses below are those from which copies of the syllabuses and past examination papers can be ordered.

Associated Examining Board
Stag Hill House
Guildford
Surrey GU2 5XJ

Northern Examinations and Assessment Board
Manchester M15 6EU

University of Cambridge Local Examinations Syndicate
Syndicate Buildings
1 Hills Road
Cambridge CB1 2EU

University of London Examinations and Assessment Council
Stewart House
32 Russell Square
London WC1B 5DN

Welsh Joint Examinations Committee
245 Western Avenue
Cardiff CF5 2YX

GENERAL EXAMINATION HINTS

1 **Read the questions carefully**, especially all the numbers, signs and indices.

2 **Choose your questions carefully**; in particular, start with a question you can do. It is most unlikely that you have to start with Q.1.

3 In short-answer papers, **if you cannot see how to do a question** straightaway, **leave it out for the time being** and return later if you have time. You do not have to answer all the questions in the paper to get quite a good grade.

4 Always **make a rough estimate of any calculation**. If you are using a calculator, set out each calculation clearly, so that you – and the examiner – can see what you are trying to do.

5 Always try to **keep solutions as simple as possible** – if you find a difficult method, you may make a mistake in following it. In particular, integrals are often simpler than they appear at first sight!

6 **Do not cross out a solution because you think it is wrong**. Part of your solution may well be correct, and you may lose the marks for this if you have crossed it out.

7 If you have finished before the end of the examination, **check your work carefully**.

8 If you are 'stuck' in any one question, **check that you have used all the information given**, and see whether you can get any ideas from an earlier part of the question.

9 Where a question says 'Hence or otherwise', **it is almost always easier to try 'hence'**, using the hint given.

10 Formula booklets are provided in all examinations. Do make sure that you are **familiar** with these **formulae booklets**, and ensure that you are in the habit of using them constantly.

1 FUNCTIONS

1.1 Notes
1.2 Composite functions
1.3 Composite functions; range; inverse function
1.4 Range; inverse function; composite function; points of inflexion
1.5 Sketch; inverse function; composite function; one-to-one function
1.6 Sketch of modulus function; inequality
1.7 Deducing sketch of $y = f(x-1)$ and $y = 2f(\frac{1}{2}x)$ from $y = f(x)$

1.1 Notes

A **function** f maps one element in a set (the domain) into one and only one element in another set (the range). The inverse function f^{-1} maps this image back into the original element.

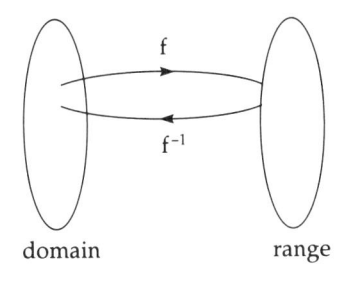

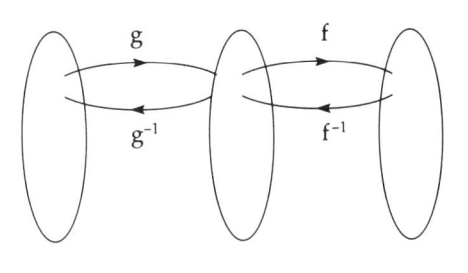

The **composite function** fg (sometimes written f∘g) means 'first find the image under g, then the image of that under f'. The inverse of the function fg is $g^{-1} f^{-1}$.
An **even function** f is such that $f(x) = f(-x)$; its graph is symmetrical about Oy. Examples of even functions are

$f : x \to x^2$, $f : x \to \cos x$, $f : x \to e^{-x^2}$.

An **odd function** f is such that $f(-x) = -f(x)$; its graph is symmetrical about the origin. Examples of odd functions are

$f : x \to x$, $f : x \to \sin x$, $f : x \to 1/x$.

A **periodic function** f is such that $f(x) = f(x+a)$. If a is the smallest number for which this is true, the function f has period a. Examples of periodic functions are

$f : x \to \sin x$, period 2π, since $\sin(x) = \sin(x + 2\pi) = \sin(x + 4\pi)\ldots$
$f : x \to \tan x$, period π, since $\tan x = \tan(x + \pi) = \tan(x + 2\pi)\ldots$

MATHEMATICS REVISION WORKBOOK

Relationship between graphs of functions

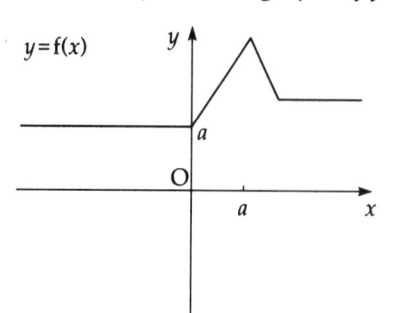

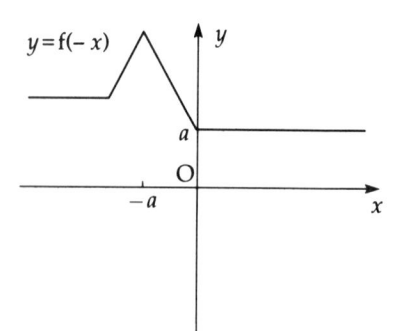

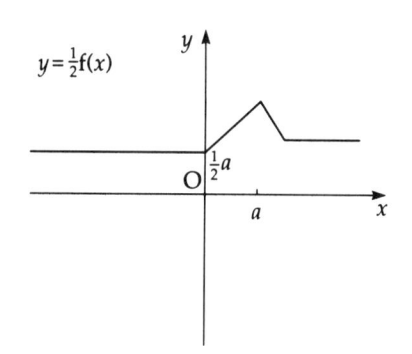

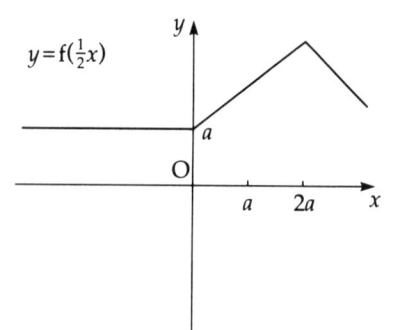

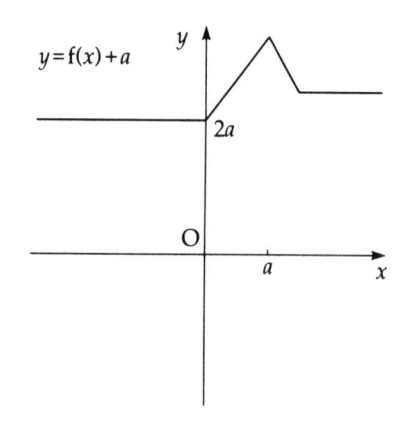

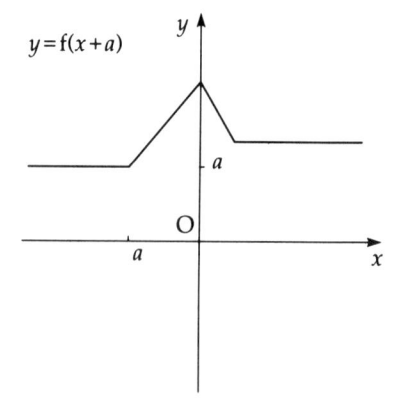

$y = f(x)$

$y = f(-x)$

$y = \frac{1}{2}f(x)$

$y = f(\frac{1}{2}x)$

$y = f(x) + a$

$y = f(x+a)$

FUNCTIONS

1.2 Composite functions

Question
The functions g and h are defined by
$$g : x \mapsto 3x^2 + 2, \quad x \in \mathbb{R},$$
$$h : x \mapsto \sqrt{\left(\frac{x-2}{3}\right)} \quad x \geq 2.$$

Find hg(x) for

(i) $x \geq 0$,
(ii) $x < 0$. (C)

Solution
Since $g : x \mapsto 3x^2 + 2$, to find hg(x) we first need
$$\sqrt{\left(\frac{(3x^2+2)-2}{3}\right)} = \sqrt{(x^2)}.$$

If $x \geq 0$, $\sqrt{(x^2)} = x$; if $x < 0$, $\sqrt{(x^2)} = -x$

so the solutions are

(i) $\underline{hg(x) = x}$,
(ii) $\underline{hg(x) = -x}$.

Comments
The notation hg(x) is equivalent to $h \circ g(x)$, and means 'first apply function g, then apply function h'.
h is only defined when $x \geq 2$, but $g(x) = 3x^2 + 2 \geq 2$ for all real x, so we do not have to consider this restriction separately.

1.3 Composite functions; range; inverse function

Question
The functions f and g are defined by
$$f : x \mapsto x^2 + 3, \ x \in \mathbb{R},$$
$$g : x \mapsto 2x + 1, \ x \in \mathbb{R}.$$

(a) Find, in a similar form, the function fg.
(b) Find the range of the function fg.
(c) Solve the equation
$$f(x) = 12 \, g^{-1}(x).$$ (L)

3

MATHEMATICS REVISION WORKBOOK

Solution

(a) Apply the function g first, so
$$fg : x \mapsto (2x+1)^2 + 3.$$
Now $(2x+1)^2 + 3 = 4x^2 + 4x + 1 + 3$
$$= 4(x^2 + x + 1)$$
so $\underline{fg : x \mapsto 4(x^2 + x + 1).}$

(b) To find the range of the function fg, it is easier to use the form
$(2x+1)^2 + 3$, for $(2x+1)^2 \geq 0$, $x \in \mathbb{R}$
so $\qquad (2x+1)^2 + 3 \geq 3, x \in \mathbb{R}$
and the range y is given by $\underline{y \geq 3}$.

(c) Since $\qquad g : x \mapsto 2x + 1,$
$\qquad g^{-1} : x \mapsto \tfrac{1}{2}(x-1)$
So $\qquad f(x) = 12\, g^{-1}(x)$
$\Rightarrow x^2 + 3 = 12 \times \tfrac{1}{2}(x-1)$
$\Rightarrow x^2 + 3 = 6(x-1)$
$\Rightarrow x^2 - 6x + 9 = 0$
$\Rightarrow (x-3)^2 = 0$
$\Rightarrow \underline{x = 3}.$

Comments

- To find the inverse function, if this cannot be done by inspection, denote the image by y, here
$$y = 2x + 1,$$
so
$$y - 1 = 2x,$$
$$\underline{x = \tfrac{1}{2}(y-1)}.$$

•• An alternative solution is
$$f(x) = 12\, g^{-1}(x)$$
$\Rightarrow \qquad \dfrac{1}{12} f(x) = g^{-1}(x)$
$\Rightarrow \qquad g\left[\dfrac{1}{12} f(x)\right] = x$
ie $\qquad g\left[\dfrac{x^2 + 3}{12}\right] = x$

4

ie
$$2\left[\frac{x^2+3}{12}\right]+1 = x$$

$$\frac{x^2+3}{6}+1 = x,$$

$$x^2+3+6 = 6x,$$
$$x^2-6x+9 = 0, \text{ as before.}$$

1.4 Range; inverse function; composite function; points of inflexion

Question

The functions f, g are defined by

$$f(x) = x^2 + 1 \ (x \geq 0)$$
and
$$g(x) = e^{-x} \ (x \geq 0).$$

(a) State the ranges of f and g.
(b) Find expressions for $f^{-1}(x)$ and $g^{-1}(x)$.
(c) Find an expression for $f \circ g(x)$ and hence solve the equation
$$3f \circ g(x) = 4g(x) + 2.$$
(d)

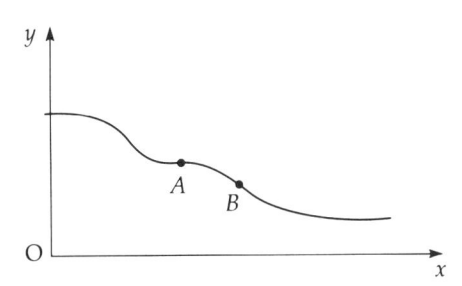

The figure shows a sketch graph of the function h defined by
$$h(x) = f(x)g(x).$$
Find the coordinates of the points of inflexion A and B. (WJEC)

Hint A sketch graph of f and of g will help us get started.

MATHEMATICS REVISION WORKBOOK

Solution

(a)

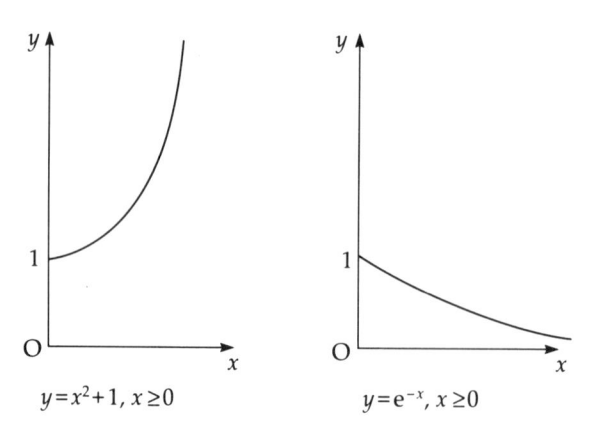

$y = x^2 + 1,\ x \geq 0$ $\qquad$ $y = e^{-x},\ x \geq 0$

Since $x^2 \geq 0$, $x^2 + 1 \geq 1$, so the range of f is all values greater than or equal to 1; we can write this

$$f(x) \geq 1.$$

Since e^{-x} decreases from 1, but is never negative, the range of g is all values from 1 down to but not including 0, which we can write as

$$0 < g(x) \leq 1.$$

(b) Denoting the image of x under f by y,

$$y = x^2 + 1$$
$$\therefore \quad x^2 = y - 1$$
$$\therefore \quad x = +\sqrt{(y-1)}, \text{ taking only the positive}$$

square root as the function is only defined over the domain $x \geq 0$.
With the usual notation, we write

$$f^{-1}(x) = +\sqrt{(x-1)}.$$

Similarly with g. Denoting the image under g by y,

$$y = e^{-x}$$
$$\therefore \quad \ln y = -x,$$
$$x = -\ln y,$$

and so
$$g^{-1}(x) = -\ln x.$$

(c) fg(x) means 'first g then f',

so
$$f \circ g(x) = [e^{-x}]^2 + 1$$
$$= e^{-2x} + 1.$$

Substituting in the equation
$$3f \circ g(x) = 4g(x) + 2$$
$$3[e^{-2x} + 1] = 4e^{-x} + 2$$
ie
$$3e^{-2x} - 4e^{-x} + 1 = 0.$$

This is a quadratic in e^{-x}, so write $z = e^{-x}$, and the equation becomes
$$3z^2 - 4z + 1 = 0,$$
$$(3z - 1)(z - 1) = 0,$$
$$z = \tfrac{1}{3} \text{ or } 1.$$

If $z = \tfrac{1}{3}$,
$$e^{-x} = \tfrac{1}{3}$$
$$-x = \ln(\tfrac{1}{3}),$$
$$= -\ln 3$$
∴
$$x = \ln 3$$

If $z = 1$,
$$e^{-x} = 1,$$
$$-x = \ln 1$$
∴
$$x = 0$$

so
$$\underline{\underline{x = 0 \text{ or } \ln 3}}$$

are the solutions of the equation.

(d) Points of inflexion occur when the second derivative is zero and changes sign, so we require $\dfrac{d^2 h}{dx^2}$.

Since
$$h(x) = (x^2 + 1)\, e^{-x}$$
$$\frac{dh}{dx} = 2xe^{-x} + (x^2 + 1)(-e^{-x})$$
$$= e^{-x}(-x^2 + 2x - 1)$$
$$= -e^{-x}(x^2 - 2x + 1)$$
$$= -e^{-x}(x - 1)^2$$

and
$$\frac{d^2h}{dx^2} = e^{-x}(x-1)^2 - e^{-x}[2(x-1)]$$
$$= e^{-x}[x^2 - 2x + 1 - 2x + 2]$$
$$= e^{-x}[x^2 - 4x + 3]$$
$$= e^{-x}(x-1)(x-3). \tag{1}$$

Since e^{-x} is never equal to 0, $d^2h/dx^2 = 0$ when $x = 1$ or $x = 3$. Although no scale is given on the axes, it looks as though A is when $x = 1$ and B when $x = 3$. From (1), we see that as x goes through the value $x = 1$, d^2h/dx^2 changes sign, and again as x goes through the value $x = 3$, d^2h/dx^2 changes sign, so that both these points $x = 1$, $x = 3$ are points of inflexion. To find the y coordinates, the equation of the curve is $y = (x^2+1)e^{-x}$, so when $x = 1$, $y = 2e^{-1}$, when $x = 3$, $y = 10e^{-3}$. The coordinates of A are $(1, 2e^{-1})$, of B are $(3, 10e^{-3})$.

1.5 Sketch; inverse function; composite function; one-to-one function

Question

(a) Functions g and h are defined by
$$g: x \mapsto \ln x, \quad x \in \mathbb{R}, x > 0$$
$$h: x \mapsto 1 + x, \quad x \in \mathbb{R}.$$
The function f is defined by
$$f: x \mapsto gh(x), \quad x \mapsto x > -1.$$
(i) Sketch the graph of $y = f(x)$
(ii) Write down expressions for $g^{-1}(x)$ and $h^{-1}(x)$.
(iii) Write down an expression for $g^{-1}h^{-1}(x)$.
(iv) Sketch the graph of $y = g^{-1}h^{-1}(x)$.
(b) The function q is defined by
$$q: x \mapsto x^2 - 4x, \quad x \in \mathbb{R}, |x| < 1.$$
Show, by means of a graphical argument or otherwise, that q is one-one, and find an expression for $q^{-1}(x)$. (C)

Hint $f: x \mapsto gh(x)$ means 'first h, then g', so $f(x) = \ln(1+x)$.

Solution

(a)(i) Since $f: x \mapsto gh(x)$, $f: x \mapsto \ln(1+x)$. The graph is given below:

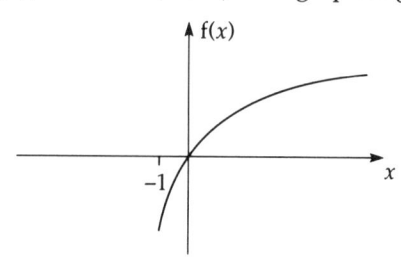

FUNCTIONS

(ii) $g^{-1}(x) : x \mapsto e^x$, $h^{-1}(x) : x \mapsto x - 1$
(iii) $g^{-1}h^{-1}(x) = g^{-1}(x-1) = e^{x-1}$; $g^{-1}h^{-1}(x) : x \mapsto e^{x-1}$
(iv) The graph of $y = e^{x-1}$ is given below:

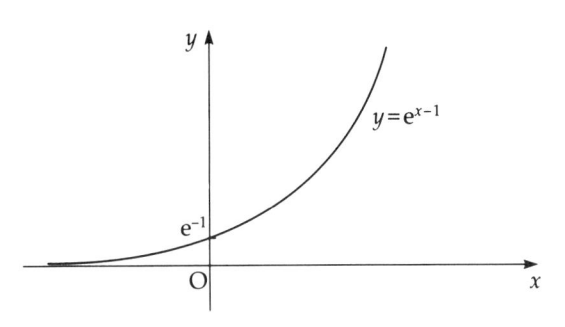

(b)

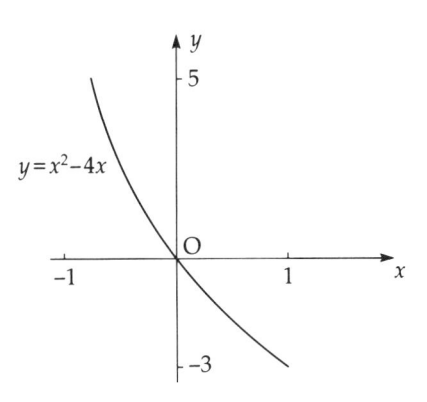

The function q is only defined over the interval $|x| < 1$, ie $-1 < x < 1$.
Over this interval, we can see that one value of x corresponds to one and only one value of y; one value of y corresponds to one and only one value of x, so q is one-one.

Since
$$q(x) = x^2 - 4x, \ q(x) = x^2 - 4x + 4 - 4$$
$$= (x-2)^2 - 4,$$
$$q(x) + 4 = (x-2)^2$$
$$x = 2 - \sqrt{(q(x) + 4)}$$
ie
$$q^{-1}(x) : x \mapsto 2 - \sqrt{(x+4)}.$$

Comments
- The graph of q(x) is only part of the graph of $y = x^2 - 4x$, as q is only defined over the

interval $-1 < x < 1$. The graph below shows $y = q(x)$ as a continuous line; that of $y = x^2 - 4x$ as a hatched line:

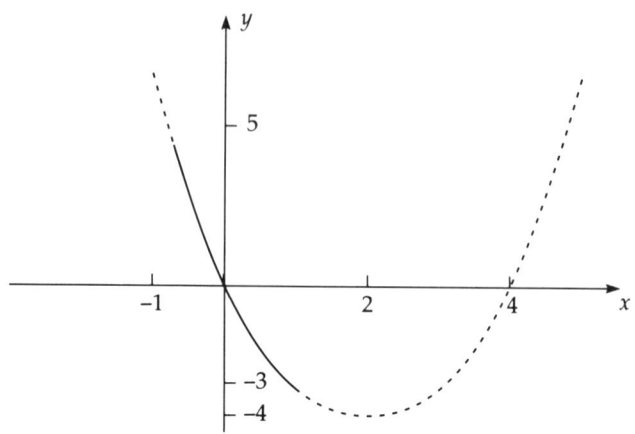

- - Since $q(x)$ is only defined over the interval $-1 < x < 1$, we want the negative square root; $2 + \sqrt{(y+4)}$ would give the part of the curve to the right of $x = 2$.

1.6 Sketch of modulus function; inequality

Question

(a) Sketch the graphs of $y = |x - 8|$ and $y = 8x$ using the same pair of axes.
(b) Determine the set of values of x for which $|x - 8| > 8x$.

(L)

Hint The graph of $y = x - 8$ is

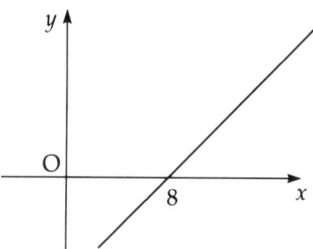

so the graph of $y = |x - 8|$ is

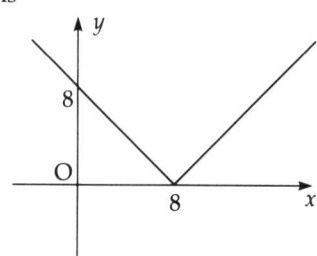

FUNCTIONS

Solution

(a) The graphs of $y = |x - 8|$ and $y = 8x$ are shown below:

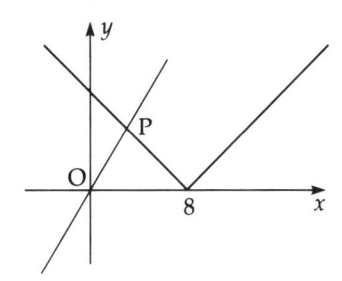

(b) From the graphs,
when $x < 8$, $y = |x - 8| \Rightarrow y = 8 - x$, so X, the x coordinate of P, is given by
$8 - X = 8X$,
$X = 8/9$, and
$|x - 8| > 8x$ when $x < 8/9$.

Comment

Check: when $x = 1$, $|x - 8| \not> 8x$.

1.7 Deducing sketch of $y = f(x - 1)$ and $y = 2f(\tfrac{1}{2}x)$ from $y = f(x)$

Question

The diagram shows a sketch of the curve $y = f(x)$ where $f(x) = 0$ for $x \leq 1$ and $x \geq 4$:

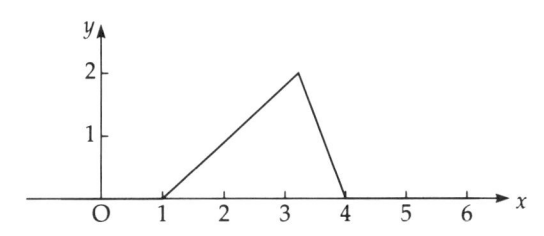

Sketch, on separate axes, the graph of the curves with equations
(a) $y = f(x - 1)$,
(b) $y = 2f(\tfrac{1}{2}x)$. (AEB)

Solution

(a) Looking at a few specific values for $y = f(x - 1)$,
when $x = 1$, $y = f(0)$,

11

when $x = 2$, $y = f(1)$,
when $x = 3$, $y = f(2)$,
so the graph has been moved 1 unit to the right.

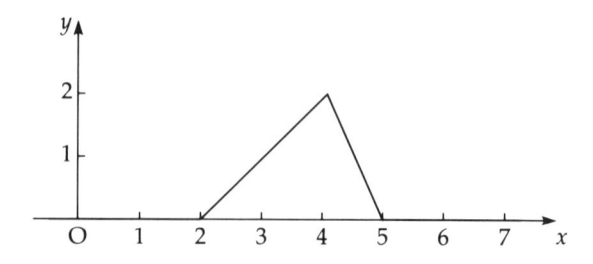

(b) Again taking some specific values of x, for $y = 2f(\frac{1}{2}x)$,
when $x = 1$, $y = 2f(\frac{1}{2}) = 0$
when $x = 2$, $y = 2f(1) = 0$
when $x = 3$, $y = 2f(1.5)$
when $x = 4$, $y = 2f(2)$,
when $x = 6$, $y = 2f(3)$, the greatest value,
when $x = 8$, $y = 2f(4) = 0$.

So the range on the x-axis has been doubled, and the range on the y-axis has also been doubled. The factor 2 outside $f(x)$ shows the range on the y-axis is doubled; the $(\frac{1}{2}x)$ shows that for $2 \leq x \leq 8$, $1 \leq \frac{1}{2}x \leq 4$.

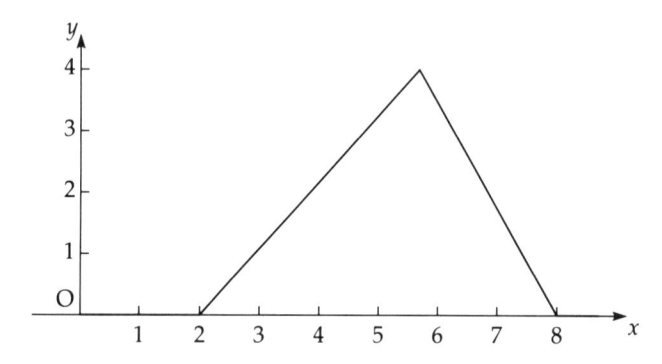

2 POLYNOMIALS

2.1 Notes
2.2 Factor theorem; factorising
2.3 Factor theorem; inequality
2.4 Remainder theorem
2.5 Factor theorem; solving a cubic by factorising

2.1 Notes
Some useful factors

$$x^2 - y^2 = (x-y)(x+y)$$
$$\pi r^2 h - \pi R^2 h = \pi h(r-R)(r+R)$$
$$x^3 - y^3 = (x-y)(x^2+xy+y^2)$$
$$x^3 + y^3 = (x+y)(x^2-xy+y^2).$$

Products

$$(a+b)(a-b) = a^2 - b^2$$
$$(a+b)^2 = a^2 + 2ab + b^2$$
$$(a-b)^2 = a^2 - 2ab + b^2.$$

The factor theorem

When a polynomial $f(x)$ is such that $f(a) = 0$, $(x-a)$ is a factor of $f(x)$.
Conversely, if $(x-a)$ is a factor of $f(x)$, $f(a) = 0$.
When $(x-a)^2$ is a factor of $f(x)$, so that $(x-a)$ is a repeated factor, then $f(a) = 0$ and $f'(a) = 0$.

The remainder theorem

When a polyomial $f(x)$ is divided by $(x-a)$, the remainder is $f(a)$.

2.2 Factor theorem; factorising
Question

Given that $f(x) = 2x^3 + 5x^2 - 4x - 3$,

MATHEMATICS REVISION WORKBOOK

(i) find whether
   (a) $x - 2$,
   (b) $x + 3$,
is a factor of $f(x)$;
(ii) find all the linear factors of $f(x)$.

Solution

(i) By the factor theorem, $(x - a)$ is a factor of $f(x)$ if $f(a) = 0$.
Now, $f(2) = 2(2^3) + 5(2^2) - 4(2) - 3$
$= 25$, so $\underline{(x - 2) \text{ is not a factor of } f(x)}$,
but $f(-3) = 2(-3)^3 + 5(-3)^2 - 4(-3) - 3$
$= 0$, so $\underline{(x + 3) \text{ is a factor of } f(x)}$.

(ii) To factorise $f(x)$ completely, as we know one factor $(x + 3)$, we write
$$2x^3 + 5x^2 - 4x - 3 = (x + 3)(\ldots\ldots\ldots\ldots).$$

where the bracket contains a quadratic expression in x.
Looking at the coefficient of x^3 on both sides of the equation, we have $2x^3$ on the left hand side, so the bracket must begin with $2x^2$; since we have a constant term -3 on the left hand side, the bracket must end with the term -1, giving
$$2x^3 + 5x^2 - 4x - 3 = (x + 3)(2x^2 \ldots\ldots - 1).$$

To find the term in x, on the left hand side we have $5x^2$,
   on the right hand side so far we have $6x^2$,
so that the missing term must be $-x$,
and
$$2x^3 + 5x^2 - 4x - 3 = (x + 3)(2x^2 - x - 1).$$

To factorise the quadratic, the only possible factors of $2x^2$ are $2x$ and x, of -1 are -1 and $+1$, so we have
$$\underline{2x^3 + 5x^2 - 4x - 3 = (x + 3)(x - 1)(2x + 1).}$$

Comment

As an alternative method, we could have continued to use the factor theorem and found $f(1) = 0$, giving a factor of $(x - 1)$, so
$$2x^3 + 5x^2 - 4x - 3 = (x + 3)(x - 1)(\ldots\ldots).$$

Looking at the terms in x^3, the bracket on the right hand side must begin with the term $2x$; looking at the constant terms, the bracket must end with the term 1, so the factors of $f(x)$ are
$$(x + 3), (x - 1) \text{ and } (2x + 1).$$

POLYNOMIALS

Notice how easy and useful it was to consider only the terms of highest and of lowest degree. We could see $2x^3 = x \times 2x^2$,
$$-3 = 3 \times (-1).$$

2.3 Factor theorem; inequality

Question

(a) Show that $(x - 2)$ is a factor of $x^3 - 9x^2 + 26x - 24$.
(b) Find the set of values of x for which $x^3 - 9x^2 + 26x - 24 < 0$.

(AEB)

Solution

(a) By the factor theorem, if $f(x) \equiv x^3 - 9x^2 + 26x - 24$, $(x - 2)$ is a factor of $f(x)$ if $f(2) = 0$.

Now $f(2) = 2^3 - 9(2^2) + 26 \times 2 - 24$
$= 0$,
so $(x - 2)$ is a factor of $x^3 - 9x^2 + 26x - 24$.

(b) Knowing $x - 2$ is a factor, we write
$$x^3 - 9x^2 + 26x - 24 = (x - 2)(\ldots\ldots\ldots\ldots).$$
The first term in the bracket must be x^2, to give the correct term in x^3; the last term must be $+12$, to give the constant term -24, so we have
$$x^3 - 9x^2 + 26x - 24 = (x - 2)(x^2 \ldots\ldots + 12).$$
Looking at the term in x^2, we have $-9x^2$ on the left hand side; we already have $-2x^2$ on the right hand side, so the missing term in the bracket must be $-7x$, and
$$x^3 - 9x^2 + 26x - 24 = (x - 2)(x^2 - 7x + 12)$$
$$= (x - 2)(x - 3)(x - 4).$$
so $f(x) = 0$ if $(x - 2)(x - 3)(x - 4) = 0$, ie $x = 2, 3$ or 4.
The rough shape of the graph of $y = x^3 + ax^2 + bx + c$ is given below.

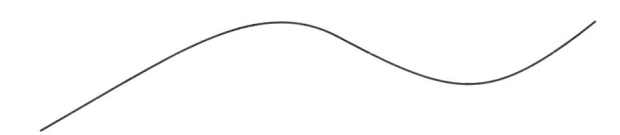

so that since $f(x) = 0$ when $x = 2, 3$ or 4, the graph of $y = f(x)$ is:

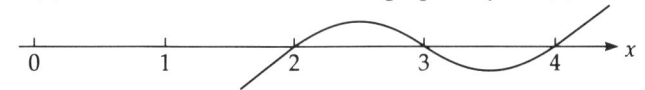

and $f(x) < 0$ if $x < 2$ or $3 < x < 4$.

15

MATHEMATICS REVISION WORKBOOK

Comments

Using a graphics calculator we could save a lot of this work. The graph of $y = x^3 - 9x^2 + 26x - 24$ shows that it crosses the x-axis at (or at any rate, very near), $x = 2, 3$ or 4. By the factor theorem we can show that $f(3) = f(4) = 0$, so that the roots of $f(x) = 0$ are exactly $x = 2, 3$ and 4. The graph then shows us that $f(x) < 0$ if $x < 2$ or $3 < x < 4$.

2.4 Remainder theorem

Question

A quadratic polynomial $ax^2 + bx + c$ leaves remainder 1 when divided by $(x - 2)$ and remainder 2 when divided by $(x - 1)$. If

$$ax^2 + bx + c \equiv a(x - 1)(x - 2) + px + q,$$

find the numerical values of p and q.

(O & C)

Solution

Using the remainder theorem, denoting $ax^2 + bc + c$ by $f(x)$, since division by $(x - 2)$ leaves remainder 1, we have

$$f(2) = 1,$$

ie $\qquad 4a + 2b + c = 1.$

Similarly $\qquad f(1) = 2,$

ie $\qquad a + b + c = 2.$

We do not have enough equations to find a, b and c, but we do not need to do so, for we have to find only p and q.
Substituting $x = 2$ in $ax^2 + bx + c = a(x - 1)(x - 2) + px + q$, we have

$$4a + 2b + c = 2p + q,$$

and substituting $x = 1$, we have

$$a + b + c = p + q.$$

We have already found the values of $ax^2 + bx + c$ when $x = 2$ and when $x = 1$

so $\qquad 1 = 2p + q,$
$\qquad\qquad 2 = p + q.$

Solving, $\qquad\qquad \underline{p = -1 \text{ and } q = 3}.$

POLYNOMIALS

Note

This illustrates well the need to read the question carefully and to answer exactly what the question asks.

2.5 Factor theorem; solving a cubic by factorising

Question

The polynomial
$$p(x) \equiv 2x^3 - 9x^2 + kx - 13,$$
where k is a constant, has $(2x - 1)$ as a factor.

(a) Determine the value of k.
(b) Find the three roots, real and complex, of the equation $p(x) = 0$.

(AEB)

Solution

(a) Since $(2x - 1)$ is a factor of $p(x)$, $p(\tfrac{1}{2}) = 0$,

ie $\qquad 2(\tfrac{1}{2})^3 - 9(\tfrac{1}{2})^2 + k(\tfrac{1}{2}) - 13 = 0$
ie $\qquad \tfrac{1}{4} - 2\tfrac{1}{4} + \tfrac{1}{2}k - 13 = 0,$
$\qquad \tfrac{1}{2}k = 15,$
$\qquad \underline{k = 30.}$

(b) Now we know $p(x) \equiv 2x^3 - 9x^2 + 30x - 13$, and that $(2x - 1)$ is a factor,
so $\qquad 2x^3 - 9x^2 + 30x - 13 = (2x - 1)(\ldots\ldots\ldots\ldots).$
Looking at the term $2x^3$, the second bracket must begin with x^2; looking at the term -13, the second bracket must end with $+13$, so we have
$$2x^3 - 9x^2 + 30x - 13 = (2x - 1)(x^2 \ldots\ldots + 13).$$
Look at the term in x^2; on the left hand side of the equation we have $-9x^2$; on the right hand side we have so far $-x^2$, so the term in x in the bracket must be $-4x$, and
$$2x^3 - 9x^2 + 30x - 13 = (2x - 1)(x^2 - 4x + 13).$$
Thus $p(x) = 0 \Rightarrow (2x - 1)(x^2 - 4x + 13) = 0,$
i.e. either $\qquad 2x - 1 = 0 \text{ or } x^2 - 4x + 13 = 0.$
If $2x - 1 = 0$, $x = \tfrac{1}{2}$.
If $x^2 - 4x + 13 = 0$, the words 'real or complex' in the question suggest that these roots may be complex, so

17

MATHEMATICS REVISION WORKBOOK

$$\begin{aligned}
& x^2 - 4x + 13 = 0 \\
\Rightarrow \quad & x^2 - 4x + 4 = -9, \text{ completing the square,} \\
& (x - 2)^2 = -9, \\
& x - 2 = 3i \text{ or } -3i, \\
& x = 2 + 3i \text{ or } 2 - 3i, \text{ so the three roots of } p(x) = 0 \text{ are} \\
& \underline{\underline{\tfrac{1}{2}, 2 + 3i \text{ or } 2 - 3i.}}
\end{aligned}$$

Notes

We could have solved equation • by using the formula to solve a quadratic equation. We could have found the quadratic factor $x^2 - 4x + 13$ by long division of $p(x)$ by $(2x - 1)$.

3 QUADRATIC EQUATIONS

3.1 Notes
3.2 Solution of a quadratic; substitution $y = \cos x$
3.3 Solution of a quadratic; substitution $y = e^x$, logarithms
3.4 Solution of two simultaneous equations; one linear and one quadratic
3.5 Condition that a quadratic equation has two real distinct roots
3.6 Sum and product of the roots of a quadratic equation
3.7 Quadratic inequalities

3.1 Notes

Solution of a quadratic equation

The roots of the equation $ax^2 + bx + c = 0$ are

$$x = \frac{-b \pm \sqrt{(b^2 - 4ac)}}{2a}.$$

The **discriminant** of this equation is $b^2 - 4ac$.

The equation has two real distinct roots if $b^2 - 4ac > 0$,
equal roots if $b^2 - 4ac = 0$,
no real roots if $b^2 - 4ac < 0$.

Sum and product of roots

If α and β are the roots of $ax^2 + bx + c = 0$,
$$\alpha + \beta = -b/a,$$
$$\alpha\beta = c/a.$$

Graph

The graph of $y = ax^2 + bx + c$ is concave upwards if a is positive, downwards if a is negative:

a positive 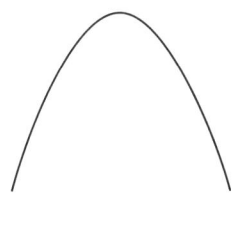   *a* negative

MATHEMATICS REVISION WORKBOOK

Inequalities

If a is positive and $b < c$, $a(x - b)(x - c) > 0 \Rightarrow x < b$ or $x > c$;
$$a(x - b)(x - c) < 0 \Rightarrow b < x < c.$$

3.2 Solution of a quadratic; substitution $y = \cos x$

Question

Solve the equation
$$9\cos^2 x - 6\cos x - 0.21 = 0, \quad 0° \leq x < 360°$$
giving each answer in degrees to 1 decimal place.

(L)

Hint Recognise this equation as a quadratic in $\cos x$, and use the formula to solve the quadratic equation.

Solution

Since $9\cos^2 x - 6\cos x - 0.21 = 0$,
$$\cos x = \frac{6 \pm \sqrt{(6^2 - 4 \times 9 \times (-0.21))}}{2 \times 9}$$
$$= \frac{6 \pm \sqrt{43.56}}{18}$$
$$= \frac{6 + 6.6}{18} \text{ or } \frac{6 - 6.6}{18}$$
$$= 0.7 \text{ or } -1/30.$$

When $\cos x = 0.7$, $x = 45.6°$, the value in the first quadrant; the value in the fourth quadrant can be seen from the graph of $y = \cos x°$ and is $(360 - 45.6)°$, ie $314.4°$:

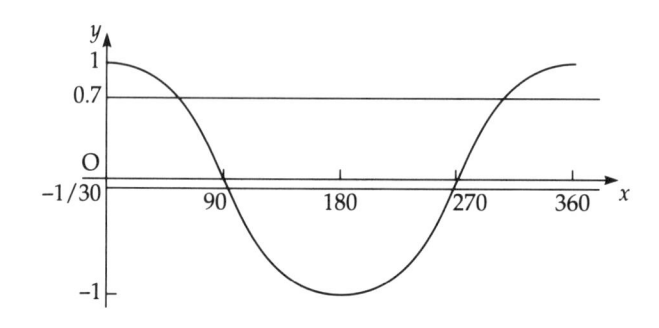

QUADRATIC EQUATIONS

When $\cos x = -1/30$, $x = 91.9°$, in the second quadrant or $268.1°$ in the third quadrant, so the required values of x are

$$\underline{\underline{45.6°, \ 91.9°, \ 268.1° \text{ and } 314.4°.}}$$

3.3 Solution of a quadratic; substitution $y = e^x$, logarithms

Question

By treating the following equation as a quadratic in e^x, find the two values of x satisfying

$$e^{2x} - 5e^x + 6 = 0.$$

(WJEC)

Solution

Write $y = e^x$, then $y^2 = e^{2x}$ and the equation becomes

$$y^2 - 5y + 6 = 0.$$

Factorising, $\quad (y-2)(y-3) = 0,$

$$y = 2 \text{ or } y = 3.$$

When $y = 2$, $e^x = 2$, so $\quad x = \ln 2$, 0.69 to 2 dp.
When $y = 3$, $e^x = 3$, so $\quad x = \ln 3$, 1.10 to 2 dp.
Thus the solutions are $\quad \underline{\underline{x = \ln 2 \text{ or } x = \ln 3.}}$

Comment

We can leave the solutions as ln 2, ln 3, as we were not required to give them to any required degree of accuracy, or any reasonable approximation would be acceptable. It often helps to check if we look at numerical values, eg ln 1 = 0, so if we had thought that ln 1 was a solution (perhaps having factorised the quadratic wrongly), we would have seen that e^0 (i.e. $y = 1$) could not be a solution of the equation.

3.4 Solution of two simultaneous equations; one linear and one quadratic

Question

Solve the simultaneous equations

$$4x^2 + y^2 = 25 \quad xy = 6$$

giving all possible pairs of values of x and y.

(WJEC)

Solution

From the second equation, $y = 6/x$, so substituting in the first equation,

21

MATHEMATICS REVISION WORKBOOK

ie
$$4x^2 + (6/x)^2 = 25,$$

ie
$$4x^2 + \frac{36}{x^2} = 25$$

$$4x^4 - 25x^2 + 36 = 0.$$

Factorising,
$$(4x^2 - 9)(x^2 - 4) = 0$$

$$4x^2 = 9 \text{ or } x^2 = 4.$$

From these we have $x = \pm 3/2$ or $x = \pm 2$.

When $x = 2$, $y = 6 \div 2$, ie 3; when $x = -2$, $y = -3$;
when $x = 3/2$, $y = 6 \div 3/2$, ie 4; when $x = -3/2$, $y = -4$,
so the solutions are

$x = 2, y = 3; x = -2, y = -3; x = 3/2, y = 4; x = -3/2, y = -4.$

Comments

- We recognise that the equation is a quadratic in x^2. We can write $y = x^2$, if we feel more confident solving $4y^2 - 25y + 36 = 0$.
- We can of course use the formula if we cannot see the factors,

$$x^2 = \frac{25 \pm \sqrt{(25^2 - 4 \times 4 \times 36)}}{2 \times 4}$$

$$= \frac{25 \pm \sqrt{(49)}}{8}, \text{ etc.}$$

- It is important that we pair the corresponding values of x and y, as done in this example. It is not sufficient to say

$x = \pm 2$ or $\pm 3/2$, $y = \pm 3$ or ± 4

as the value $x = 2$ is only a solution if $y = 3$, etc.

3.5 Condition that a quadratic equation has two real distinct roots

Question

Determine the range of values of k for which the quadratic equation
$$x^2 + 2kx + 4k - 3 = 0$$
has two real and distinct roots. (WJEC)

Hint Real distinct roots of a quadratic equation suggest that we shall use the discriminant $b^2 - 4ac$ of the quadratic. We expect then to solve an inequality to find a range of values of k.

Solution

The quadratic equation $ax^2 + bx + c = 0$ has real distinct roots if $b^2 - 4ac > 0$, so that
$$x^2 + 2kx + 4k - 3 = 0$$

QUADRATIC EQUATIONS

has real distinct roots if
$$(2k)^2 - 4(1)(4k-3) > 0$$
ie $\quad 4k^2 - 16k + 12 > 0$
ie $\quad k^2 - 4k + 3 > 0$
ie $\quad (k-1)(k-3) > 0.$

Looking at the graph of $y = (k-1)(k-3)$:

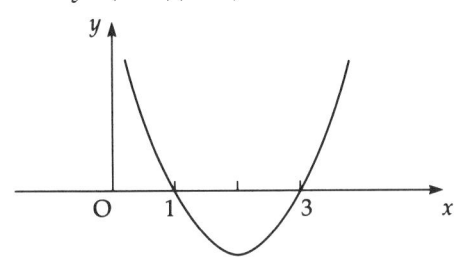

we see that $y > 0$ if $k < 1$ or $k > 3$, so that the quadratic equation $x^2 + 2kx + 4k - 3 = 0$ has real distinct roots if
$$\underline{k < 1 \text{ or } k > 3}.$$

Check

We can check these boundaries easily. When $k = 0$, we expect the quadratic to have real distinct roots, and $x^2 - 3 = 0$ has two real distinct roots, $x = \pm\sqrt{3}$. When $k = 2$, we expect the quadratic not to have real distinct roots, and for the quadratic $x^2 + 4x + 5 = 0$ the discriminant is $4^2 - 4 \times 5 = -4$, which is negative, so that quadratic does not have any real roots.

3.6 Sum and product of the roots of a quadratic equation

Question

The roots of the equation $x^2 - 2x - 4 = 0$ are α and β.

(a) Write down the values of $(\alpha + \beta)$ and $\alpha\beta$.
(b) Find the exact value of $\alpha^2 + \beta^2$.

The equation $x^2 + px + q = 0$ has roots α^2 and β^2.

(c) Find the numerical values of p and q. (L)

Solution

(a) The sum of the roots of the quadratic equation $ax^2 + bx + c = 0$ is $-b/a$; the product of these roots is c/a, so
$$\underline{\underline{\alpha + \beta = 2/1,}} \quad \text{ie} \quad \underline{\underline{2}}$$
and $\quad \underline{\underline{\alpha\beta = -4/1,}} \quad \text{ie} \quad \underline{\underline{-4}}.$

MATHEMATICS REVISION WORKBOOK

(b) Since $(\alpha + \beta)^2 = \alpha^2 + 2\alpha\beta + \beta^2$,
$$\alpha^2 + \beta^2 = (\alpha + \beta)^2 - 2\alpha\beta,$$
$$\alpha^2 + \beta^2 = (2)^2 - 2(-4),$$
$$= 12.$$

(c) Since α^2 and β^2 are the roots of $x^2 + px + q = 0$,
$$\alpha^2 + \beta^2 = -p \text{ and } \alpha^2\beta^2 = q,$$
ie $\quad p = -12 \text{ and } q = (\alpha\beta)^2 = (-4)^2 = 16$
$$\underline{p = -12 \text{ and } q = 16.}$$

Comment

If we had used the formula to solve the quadratic $x^2 - 2x - 4 = 0$ we should have found that the roots were
$$x = \frac{2 \pm \sqrt{(2^2 - 4 \times (-4))}}{2}$$
$$= 1 \pm \sqrt{5}, \text{ about } 3.24 \text{ and } -1.24.$$

Thus $\alpha^2 + \beta^2$ is approximately $(3.24)^2 + (-1.24)^2$, about 12.035, and $\alpha^2\beta^2 = (\alpha\beta)^2 = (3.24 \times -1.24)^2 = 16.14$, both close to the exact values we have obtained. If we try to use this method to obtain exact values for $\alpha^2 + \beta^2$ and $\alpha^2\beta^2$, we have to work with
$$\alpha = 1 + \sqrt{5} \text{ and } \beta = 1 - \sqrt{5}$$
and simplify expressions like $(1 + \sqrt{5})^2$, $(1 - \sqrt{5})^2$ and $(1 + \sqrt{5})(1 - \sqrt{5})$. It is easier to use the sums and products of the roots.

3.7 Quadratic inequalities

Question

Find the set of values of x for which
$$2x(x + 3) > (x + 2)(x - 3).$$
(L)

Solution

$\Rightarrow \qquad 2x(x + 3) > (x + 2)(x - 3)$
$\Rightarrow \qquad 2x^2 + 6x > x^2 - x - 6$
$\Rightarrow \qquad x^2 + 7x + 6 > 0$
$\qquad \qquad (x + 6)(x + 1) > 0.$

QUADRATIC EQUATIONS

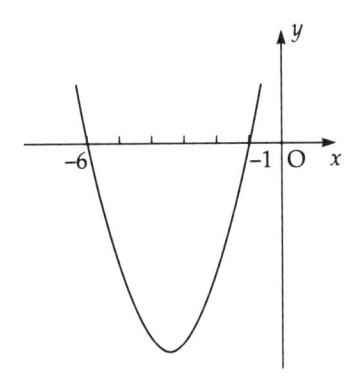

$x^2 + 7x + 6 = 0$ when $(x+6)(x+1) = 0$
ie $\quad x = -6$ or $x = -1$.
From the graph, $(x+6)(x+1) > 0$ when $\underline{\underline{x < -6 \text{ or } x > -1}}$.

(L)

Comment

We can check our answers by taking easy values for x, eg $x = 0$ satisfies $x > -1$ so should satisfy the original inequality.

When $x = 0$, $2x(x+3) = 0$.

When $x = 0$, $(x+2)(x-3) = -6$, and $0 > -6$, so $x = 0$ satisfies the original inequality as expected. Similarly we can show that, eg, $x = -2$ does not satisfy the original inequality, which we expect as $x = -2$ is not in either range $x < -6$ or $x > -1$.

4 ARITHMETIC SERIES, GEOMETRIC SERIES

4.1 Notes
4.2 Sum of an arithmetic progression
4.3 General term; sum of an arithmetic progression
4.4 Sum of a geometric progression
4.5 Sum of a geometric series; sum of an arithmetic series; sum to infinity of a geometric series
4.6 Sum of an arithmetic series; sum of a geometric series; sum to infinity of a geometric series
4.7 Condition for a geometric series to converge to a fixed value

4.1 Notes

Arithmetic progression (or series)

In the arithmetic sequence
$$a, (a+d), (a+2d), \ldots\ldots$$
the **nth** term is
$$a + (n-1)d$$
and the **sum** of the first n terms is
$$\tfrac{1}{2}n[2a + (n-1)d],$$
ie
$$\tfrac{1}{2}n[a + \ell], \text{ where } \ell \text{ is the last (ie } n\text{th) term.}$$

Geometric progression (or series)

In the geometric sequence
$$a, ar, ar^2, \ldots\ldots$$
the **nth** term is
$$ar^{n-1}$$
and the **sum** of the first n terms is
$$a\,\frac{1-r^n}{1-r}.$$

ARITHMETIC SERIES, GEOMETRIC SERIES

When $|r| < 1$, the sum of the series approaches a value called the **sum to infinity**, which is

$$\frac{a}{1-r}.$$

4.2 Sum of an arithmetic progression

Question

Five numbers are in an arithmetic progression and the sum of their squares is 147.5. The middle number is 5. Find the other four numbers.

(WJEC)

Hint Although we think of an A.P. as

$$a, (a+d), (a+2d), (a+3d), (a+4d)$$

since we are given the middle number in this question it will be easier to take the A.P. as

$$(5-2d), (5-d), 5, (5+d), (5+2d).$$

Solution

Taking the A.P. in the form suggested above, since the sum of the squares is 147.5

$$(5-2d)^2 + (5-d)^2 + 5^2 + (5+d)^2 + (5+2d)^2 = 147.5$$

ie $(25 - 20d + 4d^2) + (25 - 10d + d^2) + 5^2 + (25 + 10d + d^2) + (25 + 20d + 4d^2) = 147.5$

ie $125 + 10d^2 = 147.5$

$d^2 = 2.25,$

$d = \pm 1.5.$

When $d = 1.5$, the first term is $5 - 2 \times 1.5 = 2$, and the A.P. is

2, 3.5, 5, 6.5, 8.

If we consider the negative value $d = -1.5$, we obtain the same terms but in the reverse order, so there is only one solution and the other four terms are

<u>2, 3.5, 6.5 and 8.</u>

Comment

It is easy to check that the sum of all five squares is 147.5.

4.3 General term; sum of an arithmetic progression

Question

In an arithmetic progression, the 8th term is twice the 3rd term and the 20th term is 110.

(a) Find the common difference.
(b) Determine the sum of the first 100 terms.

(AEB)

MATHEMATICS REVISION WORKBOOK

Hint In the arithmetic progression the general term, the nth term, is
$$a + (n-1)d$$
and the sum of the first n terms is
$$\tfrac{1}{2}n\,[\,2a + (n-1)d\,]$$
where a is the first term and d the common difference.

Solution

(a) Since the 8th term is twice the 3rd term,
$$(a + 7d) = 2\,(a + 2d),$$
ie
$$3d = a.$$
Since the 20th term is 110,
$$a + 19d = 110,$$
ie
$$3d + 19d = 110,$$
$$\underline{\underline{d = 5.}}$$

(b) Since $d = 5$ and $a = 3d$, $a = 15$.
The sum S of the first 100 terms is given by
$$S = \tfrac{1}{2} \times 100 \times [\,2a + 99d\,]$$
$$= 50\,[30 + 297\,]$$
$$\underline{\underline{S = 16350.}}$$

4.4 Sum of a geometric progression

Question

The sum of the first and second terms of a geometric progression is 108 and the sum of the third and fourth terms is 12. Find the two possible values of the common ratio and the corresponding values of the first term.

(AEB)

Solution

Denoting the geometric series by a, ar, ar^2, ar^3 ... we have
$$a + ar = 108 \quad (1)$$
and
$$ar^2 + ar^3 = 12$$
ie
$$ar^2(1 + r) = 12. \quad (2)$$
From (1) and (2), $r^2 = \tfrac{1}{9}$, so $r = \tfrac{1}{3}$ or $-\tfrac{1}{3}$.

28

ARITHMETIC SERIES, GEOMETRIC SERIES

From (1), when $r = \frac{1}{3}$, $a = 81$; when $r = -\frac{1}{3}$, $a = 162$, so the two corresponding values of the first term are

$$\underline{81 \text{ and } 162}.$$

Comment

We can check that the series are $\quad 81, 27, 9, 3 \ldots$
and $\quad 162, -54, 18, -6$
both of which satisfy the data given.

4.5 Sum of a geometric series; sum of an arithmetic series; sum to infinity of a geometric series

Question

The first three terms of a geometric series are 1, r, and s where r is negative. Given also that 1, s and r are the first three terms of an arithmetic series, show that

$$2r^2 - r - 1 = 0.$$

Hence find the value of r.
Find the sum to infinity of the geometric series.

(AEB)

Solution

The geometric series in which the first two terms are 1, r, begins

$$1, r, r^2, r^3, \ldots\ldots$$

and as s is the third term in this series, $s = r^2$.
The arithmetic series in which the first two terms are 1, s, has common difference $(s-1)$, so has third term $s + (s-1)$, ie $2s - 1$. Since the third term of this series is r we have

$$r = 2s - 1,$$
$$r = 2r^2 - 1,$$
$$2r^2 - r - 1 = 0.$$

Factorising, $(2r + 1)(r - 1) = 0$,
so $\qquad r = -\frac{1}{2}$ or 1.
But we are told r is negative, so $r = -\frac{1}{2}$.

The sum to infinity of the geometric series $a, ar, \ldots\ldots$ is $\dfrac{a}{1-r}$

so the sum to infinity of this series, $1, -\frac{1}{2}, \left(-\frac{1}{2}\right)^2 \ldots$

is $\dfrac{1}{1-\left(-\frac{1}{2}\right)}$, ie $\underline{\underline{\tfrac{2}{3}}}$.

MATHEMATICS REVISION WORKBOOK

4.6 Sum of an arithmetic series; sum of a geometric series; sum to infinity of a geometric series

Question

(a) The sum of the first 3 terms of an arithmetic progression is 15 and the sum of the first 5 terms is 40.

  (i) Find the first term and the common difference.
  (ii) Given that the sum of the first n terms is 1552, find the value of n.

(b) S_n denotes the sum of the first n terms of a geometric progression with first term a and common ratio r, so that
$$S_n = a + ar + ar^2 + \ldots\ldots + ar^{n-1}.$$
By considering $S_n - rS_n$, show that
$$S_n = \frac{a(1-r^n)}{1-r}.$$
Deduce that, if $|r| < 1$, S_n tends to a limit as $n \to \infty$ and state this limit.

(c) A geometric series has first term a and common ratio r. Its sum to infinity is 12. The sum to infinity of the squares of the terms of this geometric series is 48. Find the values of a and r. (WJEC)

Hint In (a) we shall use the formula for the sum of n terms of an A.P. Part (b) is a standard method for finding the sum of a G.P., and (c) just requires the sum to infinity of a G.P.

Solution

(a) (i) The sum of S terms of an arithmetic progression is $\frac{1}{2}n\,[\,2a + (n-1)\,d\,]$

so $\quad\quad\quad \frac{1}{2}\,(3)\,[\,2a + (3-1)\,d\,] = 15$
and $\quad\quad\quad \frac{1}{2}\,(5)\,[2a + (5-1)\,d\,] = 40.$

Simplifying, $\quad\quad 2a + 2d = 10,$
$\quad\quad\quad\quad\quad\quad 2a + 4d = 16,$

so $\quad d = 3 \quad$ and $\quad a = 2,$
the <u>first term is 2 and the common difference is 3</u>.

(ii) Using the formula for the sum of the first n terms of an A.P.,
$$S = \tfrac{1}{2}n\,[\,2a + (n-1)\,d\,]$$
$$1552 = \tfrac{1}{2}n\,[\,2 \times 2 + (n-1) \times 3\,],\text{ since } a = 2 \text{ and } d = 3,$$
$$= \tfrac{1}{2}n\,[\,4 + 3(n-1)\,]$$
$$= \tfrac{1}{2}n\,[\,3n + 1\,]$$
$$3104 = 3n^2 + n,$$
$$3n^2 + n - 3104 = 0.$$

30

ARITHMETIC SERIES, GEOMETRIC SERIES

Using the formula for solving a quadratic equation,
$$n = \frac{-1 \pm \sqrt{(1^2 + 4 \times 3 \times 3104)}}{2 \times 3}$$
$$= \frac{-1 \pm \sqrt{(37249)}}{6}$$
$$= \frac{-1 \pm 193}{6},$$

$n = 32$ or a negative value, which is clearly irrelevant to this problem, so
$$\underline{\underline{n = 32.}}$$

(b) Since $S_n = a + ar + ar^2 + ar^3 + \ldots + ar^{n-1}$
$rS_n = ar + ar^2 + ar^3 + ar^4 + \ldots + ar^n$
Subtracting, $S_n - rS_n = a - ar^n$
ie $S_n(1 - r) = a(1 - r^n)$
$$S_n = \frac{a(1 - r^n)}{1 - r}.$$

When $|r| < 1$, r^n becomes as small as we wish if n is sufficiently large, ie $r^n \to 0$, so $(1 - r^n) \to 1$, and S_n tends to the limit $\underline{\underline{\dfrac{a}{1-r}}}$.

(c) The geometric series $a, ar, ar^2, \ldots$ has sum to infinity $a/(1-r)$, from part (b). The series formed by the squares of the terms is
$$a^2, a^2r^2, a^2r^4, \ldots$$
and this is also a geometric series, first term a^2, common ratio r^2. Since $|r| < 1$, r^2 is also less than 1, so the sum to infinity of this series also exists. From the first series,
$$\frac{a}{1 - r} = 12;$$
from the second series
$$\frac{a^2}{(1 - r^2)} = 48;$$

ie $\quad a = 12(1 - r)$ and $a^2 = 48(1 - r^2)$

ie $\quad [12(1-r)]^2 = 48(1 - r^2),$
$\quad\quad 12(1-r)^2 = 4(1-r)(1+r),$
$\quad\quad\quad 3(1-r) = (1+r),$ as $(1-r) \neq 0$
$\quad\quad\quad\quad\quad 2 = 4r,$
$$r = \tfrac{1}{2}.$$

MATHEMATICS REVISION WORKBOOK

Since $r = \frac{1}{2}$, and $a/(1-r) = 12$,
$$a/\tfrac{1}{2} = 12,$$
$$a = 6,$$
thus $\quad a = 6$ and $r = \tfrac{1}{2}$.

4.7 Condition for a geometric series to converge to a fixed value

Question

State the condition under which the infinite geometric series

$$1 + \left(\frac{2x+3}{x+1}\right) + \left(\frac{2x+3}{x+1}\right)^2 + \ldots\ldots$$

has a sum and obtain the set of values of x for which this condition holds. Assuming that this condition holds, state the sum of the series in terms of x.

(O&C)

Hints An infinite geometric series has a sum if r, the common ratio, is such that $|r| < 1$, ie $-1 < r < 1$. In this series,

$$r = \frac{2x+3}{x+1},$$

so we have to solve the inequality

$$-1 < \frac{2x+3}{x+1} < 1.$$

If we have a graphics calculator, it shows the graph of $y = \dfrac{2x+3}{x+1}$ and we can read the range of values for which $-1 < y < 1$ as $-2 < x < -1.333\ldots$ We probably suspect that the upper limit is $-4/3$.

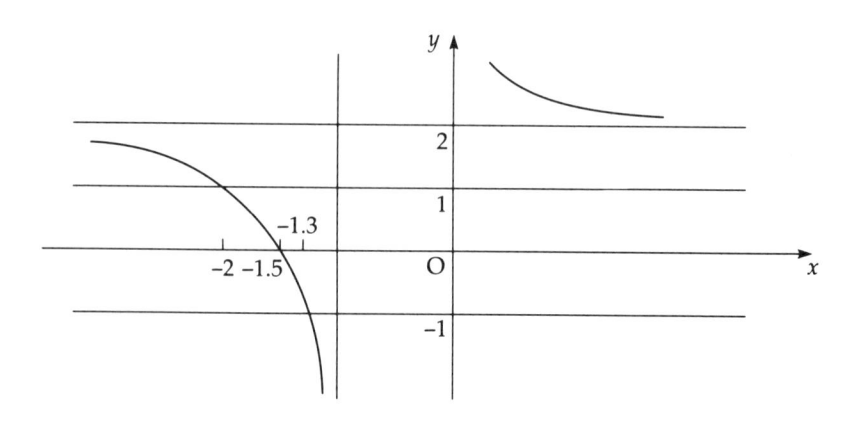

ARITHMETIC SERIES, GEOMETRIC SERIES

Solution

An infinite geometric series has a sum if $-1 < r < 1$,

ie
$$-1 < \frac{2x+3}{x+1} < 1.$$

From the graph of $y = \dfrac{2x+3}{x+1}$ on a graphics calculator,

$$-1 < y < 1 \quad \text{when} \quad -2 < x < -1.33333$$

so the geometric series has a sum when

$$\underline{\underline{-2 < x < -1.33333}}.$$

The sum of the infinite geometric series $a + ar + ar^2 + \ldots$ is $\dfrac{a}{1-r}$, so the sum S of this series is

ie
$$S = \cfrac{1}{1 - \cfrac{2x+3}{x+1}} = \underline{\underline{-\left(\frac{x+1}{x+2}\right)}}.$$

Comment

We can check this sum by choosing any value of x in the range $-2 < x < -1\tfrac{1}{3}$, say $x = -1.6$. When $x = -1.6$, $\dfrac{2x+3}{x+1} = \tfrac{1}{3}$, the series is $1 + \tfrac{1}{3} + (\tfrac{1}{3})^2 + \ldots\ldots$ and the sum is $1\tfrac{1}{2}$. Substituting in $-\left(\dfrac{x+1}{x+2}\right)$, we obtain the value $1\tfrac{1}{2}$.

If we solve the inequality by calculation, we obtain the upper limit as $-4/3$, which we probably had guessed from $-1.33\ldots\ldots$

5 INDICES AND LOGARITHMS

5.1 Notes
5.2 Positive and negative indices
5.3 Rational indices
5.4 Surds (square roots) and logarithms
5.5 Simultaneous equations involving logarithms
5.6 Simultaneous equations involving indices and logarithms

5.1 Notes

The **laws of indices** are
$$a^m \times a^n = a^{m+n}, \quad a^m \div a^n = a^{m-n}, \quad (a^m)^n = a^{mn}.$$

From these we deduce
$$a^0 = 1, \quad a^{-m} = \frac{1}{a^m} \quad \text{and} \quad a^{1/m} = \sqrt[m]{a}, \quad \text{in particular } a^{\frac{1}{2}} = \sqrt{a}.$$

Rationalising the denominator
$$\frac{1}{\sqrt{a}-\sqrt{b}} = \frac{\sqrt{a}+\sqrt{b}}{(\sqrt{a}-\sqrt{b})(\sqrt{a}+\sqrt{b})} = \frac{\sqrt{a}+\sqrt{b}}{a-b}$$

eg
$$\frac{1}{\sqrt{3}-\sqrt{2}} = \frac{\sqrt{3}+\sqrt{2}}{(\sqrt{3}-\sqrt{2})(\sqrt{3}+\sqrt{2})} = \sqrt{3}+\sqrt{2}.$$

Logarithms

From the definition of the logarithm function,
$$y = a^x \Leftrightarrow x = \log_a y,$$
where a is the **base** of the logarithm,

eg since $100 = 10^2$, $\log_{10} 100 = 2$.

From the laws of indices
$$\log x + \log y = \log(xy); \quad \log x \div \log y = \log(x/y), \quad \log(x^n) = n \log x.$$

INDICES AND LOGARITHMS

Natural (or Napierian) logarithms
Natural logarithms have base e and are written ln,
ie
$$\ln x \equiv \log_e x.$$

Changing bases
$$\log_b x = \frac{\log_a x}{\log_a b},$$ in particular $\log_{10} x = \ln x / \ln 10$, $\ln x = \log_{10} x / \log_{10} e$.

5.2 Positive and negative indices

Question
Given that $27^x = 9^{x-1}$, find the value of x.

(L)

Hint Both 27 and 9 are powers of 3, so we express the equation in powers of 3.

Solution
Since $27 = 3^3$ and 3^2,

$$27^x = 9^{x-1} \Rightarrow (3^3)^x = (3^2)^{x-1}$$

ie $\qquad 3^{3x} = 3^{2x-2}$

ie $\qquad 3x = 2x - 2$,

ie $\qquad \underline{\underline{x = -2}}.$

Comment
We can check this result easily. When $x = -2$, $27^x = 27^{-2} = \frac{1}{729}$ (or 0.00137 .... by calculator) and $9^{x-1} = 9^{-3} = \frac{1}{729}$ (or again, 0.00137 .... by calculator).

5.3 Rational indices

Question
Given that $p = t^{\frac{1}{2}} + t^{-\frac{1}{2}}$ and $q = t^{\frac{1}{2}} - t^{-\frac{1}{2}}$, find
$$p^2 q^2 + 2$$
in terms of t, giving your answer in its simplest form.

(L)

Hint We should recognise $(x+y)(x-y) = x^2 + y^2$, so that pq will be simple, and $p^2 q^2 = (pq)^2$ will be the easiest way to simplify the expression given.

35

MATHEMATICS REVISION WORKBOOK

Solution

$$pq = (t^{\frac{1}{2}} + t^{-\frac{1}{2}})(t^{\frac{1}{2}} - t^{-\frac{1}{2}})$$
$$= t - t^{-1}$$

and
$$p^2q^2 + 2 = (t - t^{-1})^2 + 2$$
$$= t^2 - 2 + t^{-2} + 2$$
$$= \underline{\underline{t^2 + t^{-2}}}, \text{ which can be written } \underline{\underline{t^2 + \frac{1}{t^2}}}.$$

Comments

- It is vital to realise that $t^{\frac{1}{2}} \times t^{\frac{1}{2}} = t$, $t^{-\frac{1}{2}} \times t^{-\frac{1}{2}} = t^{-1}$ and $t^{\frac{1}{2}} \times t^{-\frac{1}{2}} = t^0 = 1$
- Using $t \times t^{-1} = t^0 = 1$

5.4 Surds (square roots) and logarithms

Question

Express

(i) $(3 + \sqrt{2})^4$ in the form $a + b\sqrt{2}$,
(ii) $\ln(2\sqrt{e}) - \frac{1}{3} \ln(\frac{8}{e}) - \ln(\frac{e}{3})$ in the form $c + \ln d$,

where a, b, c and d are rational numbers.

(C)

(i) *Solution*

By the binomial theorem
$$(3 + \sqrt{2})^4 = 3^4 + 4(3)^3(\sqrt{2}) + 6(3)^2(\sqrt{2})^2 + 4(3)(\sqrt{2})^3 + (\sqrt{2})^4$$
$$= 81 + 108\sqrt{2} + 108 + 24\sqrt{2} + 4$$
$$= \underline{\underline{193 + 132\sqrt{2}}}.$$

(ii) *Hint* In $e = 1$, $\ln(\sqrt{e}) = \frac{1}{2}$, $\ln\left(\frac{1}{e}\right) = -1$, and, whatever the base,

$$\log ab = \log a + \log b, \ \log(a/b) = \log a - \log b.$$

Solution

$$\ln(2\sqrt{e}) - \frac{1}{3}\ln(8/e) - \ln(e/3) = \ln 2 + \ln\sqrt{e} - \frac{1}{3}[\ln 8 - \ln e] - [\ln e - \ln 3]$$
$$= \ln 2 + \frac{1}{2} - \frac{1}{3}[3 \ln 2 - 1] - [1 - \ln 3]$$
$$= \ln 2 + \frac{1}{2} - \ln 2 + \frac{1}{3} - 1 + \ln 3$$
$$= \underline{\underline{-\frac{1}{6} + \ln 3}}.$$

INDICES AND LOGARITHMS

Comments

- Remember that $(\sqrt{2})^3 = 2\sqrt{2}$, $(\sqrt{2})^4 = 4$. We can check our answer by evaluating $(3+\sqrt{2})^4$ and $193 + 132\sqrt{2}$ by calculator, obtaining 379.676 1902 .... in each case.
- • Again, check using a calculator, 0.931 945 622 ....

5.5 Simultaneous equations involving logarithms

Question

Solve the simultaneous equations
$$\log_3 x^2 y = -8$$
$$\log_3 x \cdot \log_3 y = 8.$$

(AEB)

Hint Since $\log x^2 y = 2 \log x + \log y$, the first equation looks hopeful, but there is no simpler form for the second equation.

All the logarithms are base 3, so we need not write down the base until the last stage.

Solution

From the first equation
$$2 \log x + \log y = -8 \quad (1)$$
and
$$\log x \, \log y = 8 \quad (2).$$

Multiply (1) by $\log x$, then subtract (2)

$$2 (\log x)^2 = -8 \log x - 8,$$
ie $\quad (\log x)^2 + 4 \log x + 4 = 0,$
ie $\quad (\log x + 2)^2 = 0$
ie $\quad \log_3 x = -2,$
$$x = 3^{-2},$$
$$x = \frac{1}{9}.$$

Since $\log_3 x = -2$, from (1), $\log_3 y = -4$, so $y = 3^{-4} = \frac{1}{81}$, the solutions are $\underline{\underline{x = \tfrac{1}{9}, \ y = \tfrac{1}{81}}}$.

Comment

- From the definition of a logarithm, $\log_a x = b \Leftrightarrow x = a^b$.

37

5.6 Simultaneous equations involving indices and logarithms

Question

Given the simultaneous equations
$$2^x = 3^y, \quad x + y = 1$$

show that $x = \dfrac{\ln 3}{\ln 6}$.

(C)

Solution

$$2^x = 3^y \Leftrightarrow x \ln 2 = y \ln 3, \quad \text{ie } y = x \dfrac{\ln 2}{\ln 3}.$$

Substitute in the second equation

$$x + x \dfrac{\ln 2}{\ln 3} = 1$$

ie
$$x \ln 3 + x \ln 2 = \ln 3$$
$$x (\ln 3 + \ln 2) = \ln 3$$
$$x \ln 6 = \ln 3$$
$$\underline{\underline{x = \dfrac{\ln 3}{\ln 6}}}.$$

Comments

Since $\dfrac{\log_a x}{\log_a y} = \dfrac{\log_b x}{\log_b y}$ for all bases a and b, there is no significance in the logarithms being to the base e. It was probably chosen because candidates would be more familiar with logarithms base e than with any other logarithms.

Always remember the definition of a logarithm as the inverse function of an exponent,

ie
$$a^x = b \Leftrightarrow \log_a b = x.$$

6 SUMMATION OF SERIES

6.1 Notes
6.2 Summation using formulae
6.3 Summation using identity
6.4 Summation using partial fractions

6.1 Notes
Summation of series

$$\sum_{r=1}^{n} r = \tfrac{1}{2}n(n+1),$$

$$\sum_{r=1}^{n} r(r+1) = \tfrac{1}{3}n(n+1)(n+2),$$

$$\sum_{r=1}^{n} r^2 = \tfrac{1}{6}n(n+1)(2n+1),$$

$$\sum_{r=1}^{n} r^3 = [\tfrac{1}{2}n(n+1)]^2,$$

and

$$\sum_{1}^{n} 1 = n.$$

$$\sum_{r=1}^{n}(ar^2 + br + c) = a\sum_{r=1}^{n} r^2 + b\sum_{r=1}^{n} r + c\sum_{1}^{n} 1$$
$$= a[\tfrac{1}{6}n(n+1)(2n+1)] + b[\tfrac{1}{2}n(n+1)] + cn.$$

6.2 Summation using formulae
Question
Show that

$$\sum_{r=1}^{n} r(3r+1) = n(n+1)^2.$$

Hence evaluate

$$\sum_{r=31}^{60} r(3r+1).$$

(L)

MATHEMATICS REVISION WORKBOOK

Hint Our formula booklet probably gives

$$\sum_{r=1}^{n} r^2 = \tfrac{1}{6}n(n+1)(2n+1)$$

and

$$\sum_{r=1}^{n} r = \tfrac{1}{2}n(n+1)$$

so we shall write

$$r(3r+1) = 3r^2 + r$$

and use the formulae given.

Solution

$$\begin{aligned}\sum_{r=1}^{n} r(3r+1) &= \sum_{1}^{n} 3r^2 + \sum_{1}^{n} r \\ &= 3[\tfrac{1}{6}n(n+1)(2n+1)] + \tfrac{1}{2}n(n+1) \\ &= \tfrac{1}{2}n(n+1)[2n+1+1] \\ &= \tfrac{1}{2}n(n+1)[2n+2] \\ &= \underline{\underline{n(n+1)^2}}.\end{aligned}$$

Now

$$\begin{aligned}\sum_{r=31}^{60} r(3r+1) &= \sum_{r=1}^{60} r(3r+1) - \sum_{1}^{30} r(3r+1) \\ &= 60(61)^2 - 30(31)^2 \\ &= \underline{\underline{194\ 430}}.\end{aligned}$$

6.3 Summation using identity

Question

Show that

$$3(r+1)(r+2) = (r+1)(r+2)(r+3) - r(r+1)(r+2).$$

By using this identity, or otherwise, find

$$\sum_{r=0}^{n}(r+1)(r+2)$$

giving your answer in terms of n.

(L)

40

SUMMATION OF SERIES

Solution

It is possibly easier to start with the right hand side and to factorise:
$$(r+1)(r+2)(r+3) - r(r+1)(r+2) = (r+1)(r+2)[(r+3) - r]$$
$$= (r+1)(r+2)(3)$$
$$= \underline{\underline{3(r+1)(r+2)}}.$$

From this identity we can deduce that
$$(r+1)(r+2) = \tfrac{1}{3}(r+1)(r+2)(r+3) - \tfrac{1}{3}r(r+1)(r+2)$$

and so each term in the given series can be written as the difference of two products, thus

$$\sum_{r=0}^{n} (r+1)(r+2) = 1 \times 2 + 2 \times 3 + 3 \times 4 + \ldots \ldots$$

$$= \left[\tfrac{1}{3} \times 1 \times 2 \times 3 - \tfrac{1}{3} \times 0 \times 1 \times 2\right] + \left[\tfrac{1}{3} \times 2 \times 3 \times 4 - \tfrac{1}{3} \times 1 \times 2 \times 3\right]$$
$$+ \left[\tfrac{1}{3} \times 3 \times 4 \times 5 - \tfrac{1}{3} \times 2 \times 3 \times 4\right] \ldots \ldots \left[\tfrac{1}{3} \times (n+1)(n+2)(n+3) - \tfrac{1}{3} \times n \times (n+1)(n+2)\right]$$

$$= \tfrac{1}{3} \times 1 \times 2 \times 3 - \tfrac{1}{3} \times 0 \times 1 \times 2$$
$$+ \tfrac{1}{3} \times 2 \times 3 \times 4 - \tfrac{1}{3} \times 1 \times 2 \times 3$$
$$+ \tfrac{1}{3} \times 3 \times 4 \times 5 - \tfrac{1}{3} \times 2 \times 3 \times 4$$

$$\ldots \ldots \ldots \ldots \ldots \ldots \ldots \ldots \ldots \ldots$$

$$+ \tfrac{1}{3}(n+1)(n+2)(n+3) - \tfrac{1}{3}n(n+1)(n+2)$$

$$= \tfrac{1}{3}(n+1)(n+2)(n+3),$$

all the terms cancelling, apart from the two shown, one of which is zero, so

$$\underline{\underline{\sum_{r=0}^{n} (r+1)(r+2) = \tfrac{1}{3}(n+1)(n+2)(n+3)}}.$$

Check

We can check this easily. When $n = 1$, the series is
$$1 \times 2 + 2 \times 3$$

whose sum is 8, and the formula gives $\frac{1}{3} \times 2 \times 3 \times 4$, which is also 8. The result also checks of course when $n = 0$, though the series then has only the single term 1×2.

Comments

- If we start with the left hand side of the identity, write
$$3 = (r+3) - r$$
so
$$\begin{aligned}3(r+1)(r+2) &= [(r+3) - r](r+1)(r+2) \\ &= (r+1)(r+2)(r+3) - r(r+1)(r+2).\end{aligned}$$

This is quicker but less obvious than the method shown first.

•• The method of differences is suggested by the question. If we had used the summation formulae for $\sum r^2$ and $\sum r$, we should have had to be careful to notice that the summation given

$$\sum_{r=0}^{n}(r+1)(r+2)$$

starts with $r = 0$, whereas the formulae for $\sum r^2$ and $\sum r$ start with $r = 1$, being

$$\sum_{r=1}^{n} r^2 = \tfrac{1}{6}n(n+1)(2n+1) \quad \text{and} \quad \sum_{r=1}^{n} r = \tfrac{1}{2}n(n+1).$$

6.4 Summation using partial fractions

Question

Express $\dfrac{2}{(2r-1)(2r+1)}$ in partial fractions and prove that

$$\sum_{r=1}^{n} \frac{2}{(2r-1)(2r+1)} = 1 - \frac{1}{2n+1}.$$

(O & C)

Hint There are several methods we can use to find the partial fractions. The quickest is the 'cover-up' method (see notes on 7.4), which enables us to write down the partial fractions. For other methods, refer to your textbook.

SUMMATION OF SERIES

Solution
Using the 'cover-up' method,
$$\frac{2}{(2r-1)(2r+1)} = \frac{1}{(2r-1)} - \frac{1}{(2r+1)}.$$

Now
$$\sum_{r=1}^{n} \frac{2}{(2r-1)(2r+1)} = \frac{2}{1 \times 3} + \frac{2}{3 \times 5} + \frac{2}{5 \times 7} \cdots + \frac{2}{(2n-1)(2n+1)}$$

and
$$\frac{2}{1 \times 3} = \frac{1}{1} - \frac{1}{3}$$
$$\frac{2}{3 \times 5} = \frac{1}{3} - \frac{1}{5}$$
$$\frac{2}{5 \times 7} = \frac{1}{5} - \frac{1}{7} \quad \text{etc}$$

so
$$\frac{2}{(2r-1)(2r+1)} = \left(\frac{1}{1} - \frac{1}{3}\right) + \left(\frac{1}{3} - \frac{1}{5}\right) + \left(\frac{1}{5} - \frac{1}{7}\right) \cdots$$
$$= \underline{\underline{1 - \frac{1}{2n+1}}}, \text{as all the terms have cancelled, except}$$

the first and the last.

7 PARTIAL FRACTIONS

7.1 Notes
7.2 Denominator of the form $(x-a)(x-b)$
7.3 Denominator of the form $(x-a)(x^2+b)$
7.4 Denominator of the form $x^2(x+a)$; application to integrating
7.5 Denominator of the form $(x+a)(x+b)$; expansion in series

7.1 Notes
The numerator of a **proper fraction** has degree less than the denominator,

eg $\dfrac{x}{x^2-1}$ is a proper fraction but

$\dfrac{x^2}{x^2-1}$ is an improper fraction. To express a fraction in partial fractions **make sure first that it is a proper fraction**; if it is not, then divide,

eg $\dfrac{x^2}{x^2-1} = \dfrac{x^2}{(x-1)(x+1)} = \dfrac{x^2-1+1}{(x-1)(x+1)}$

$= 1 + \dfrac{1}{(x-1)(x+1)}$

$= 1 + \dfrac{\frac{1}{2}}{x-1} - \dfrac{\frac{1}{2}}{x+1}.$

As we always have proper fractions, the degree of the numerator is always less than the degree of the denominator, so suitable forms for partial fractions are

$$\dfrac{c}{(x-a)(x-b)} = \dfrac{A}{x-a} + \dfrac{B}{x-b},$$

$$\dfrac{c}{(x-a)(x^2+b)} = \dfrac{A}{x-a} + \dfrac{Bx+C}{x^2+b},$$

$$\dfrac{c}{(x-a)(x-b)^2} = \dfrac{A}{x-a} + \dfrac{B}{x-b} + \dfrac{C}{(x-b)^2}.$$

7.2 Denominator of the form $(x-a)(x-b)$
Question
Express $\dfrac{1}{(x+1)(x-3)}$ in partial fractions.

(O & C)

PARTIAL FRACTIONS

Solution

Write
$$\frac{1}{(x+1)(x-3)} = \frac{A}{x+1} + \frac{B}{x-3}.$$

Then
$$\frac{1}{(x+1)(x-3)} = \frac{A(x-3) + B(x+1)}{(x+1)(x-3)}.$$

Equating the numerators of both fractions,
$$1 = A(x-3) + B(x+1). \tag{1}$$

As this is an identity, it is true for all values of x.

When $\quad x = 3, \quad 1 = B(3+1)$, ie $B = \frac{1}{4}$.

When $\quad x = -1, \quad 1 = A(-1-3)$, ie $A = -\frac{1}{4}$.

so
$$\frac{1}{(x+1)(x-3)} = \frac{\frac{1}{4}}{(x-3)} - \frac{\frac{1}{4}}{(x+1)}$$

altering the order of the terms so that the positive one appears first.

Comment

- The value of A is $\dfrac{1}{(-1-3)}$, which is found from the expression
$$\frac{1}{(x+1)(x-3)}$$
by covering up the factor $(x+1)$, and putting $x = -1$ in what remains. Similarly, the value of B is $\dfrac{1}{x+1}$ when x is replaced by 3, the value of x that makes the factor $(x-3)$ equal to zero.

As an alternative method, we could consider the coefficients of x in (1), and solve the two simultaneous equations
$$0 = A + B$$
$$1 = -3A + B.$$

7.3 Denominator of the form $(x - a)(x^2 + b)$

Question

Express $f(x)$ in partial fractions, where
$$f(x) = \frac{3x + 8}{(2x+1)(x^2 + 3)}. \tag{C}$$

Hint Since $f(x)$ is a proper fraction, with the degree of the numerator less than the degree of the denominator, the partial fractions in which it is to be expressed will also be proper fractions, so that the numerator of $1/(2x+1)$ will be a constant A, and the numerator of $1/(x^2+3)$ will be at most of degree one, eg $Bx+C$. In some cases we find that $B=0$.

Solution

Since $f(x)$ is a proper fraction, write

$$\frac{3x+8}{(2x+1)(x^2+3)} = \frac{A}{2x+1} + \frac{Bx+C}{x^2+3},$$

ie
$$\frac{3x+8}{(2x+1)(x^2+3)} = \frac{A(x^2+3) + (Bx+C)(2x+1)}{(2x+1)(x^2+3)}.$$

Equating the numerators,
$$3x+8 = A(x^2+3) + (2x+1)(Bx+C)$$

Since this is an identity,

the coefficients of x^2 must be equal,	$0 = A + 2B$	(1)
of x must be equal,	$3 = B + 2C$	(2)
the constants must be equal,	$8 = 3A + C$.	(3)
Subtracting twice (3) from (2)	$-13 = B - 6A$	(4)
Substituting $A = -2B$ from (1),	$-13 = B + 12B$,	
	$B = -1$.	

From (1), $A = 2$, and from (2) $C = 2$, so

$$f(x) = \frac{2}{2x+1} + \frac{-x+2}{x^2+3} \qquad \text{ie } f(x) = \frac{2}{2x+1} - \frac{x-2}{x^2+3}.$$

Check

As this is an identity, it is true for all values of x, so we can look for ones that will be easy to evaluate. When $x = 2$, $x - 2 = 0$ so that there will be only one term in the second form for $f(x)$, and $\frac{2}{2x+1} = \frac{2}{5}$ when $x = 2$, which is the value of $f(2)$.

If we want a further check, we can find A using the 'cover-up' method (see notes on 7.4).

7.4 Denominator of the form $x^2(x+a)$; application to integrating

Question

(a) Express $\dfrac{1}{x^2(2x-1)}$ in the form $\dfrac{A}{x} + \dfrac{B}{x^2} + \dfrac{C}{2x-1}$.

Hence or otherwise evaluate

$$\int_1^2 \frac{1}{x^2(2x-1)}\, dx.$$

PARTIAL FRACTIONS

(b) Find $\int x^3 (\ln 4x) \, dx$.

(c) Using the substitution $x = 3 \tan \theta$, evaluate

$$\int_0^3 \frac{1}{(9+x^2)} \, dx. \qquad \text{(AEB)}$$

Solution

(a) Using the 'cover-up' method, we can write down

$$\frac{1}{x^2(2x-1)} = \frac{A}{x} + \frac{-1}{x^2} + \frac{4}{2x-1}.$$

Multiplying by $x^2(2x-1)$,

$$1 = Ax(2x-1) - (2x-1) + 4x^2.$$

As there is no term in x^2 on the left hand side, there must be no term in x^2 on the right hand side, so $2A + 4 = 0$, $A = -2$ and

$$\frac{1}{x^2(2x-1)} = \frac{-2}{x} - \frac{1}{x^2} + \frac{4}{2x-1}.$$

Thus $\int_1^2 \frac{1}{x^2(2x-1)} \, dx = \int_1^2 \left(\frac{-2}{x} - \frac{1}{x^2} + \frac{4}{2x-1} \right) dx$

$$= \left[-2 \ln|x| + \frac{1}{x} + 2 \ln|2x-1| \right]_1^2$$

$$= (-2 \ln 2 + \tfrac{1}{2} + 2 \ln 3) - 1$$

$$= \underline{\underline{2 \ln (3/2) - \tfrac{1}{2}}}.$$

Notes

- The 'cover-up' method can be found in some A-level textbooks.[†] To find A and B, if eg

$$\frac{k}{(x-a)(x-b)} = \frac{A}{x-a} + \frac{B}{x-b}$$

cover up the factor $(x-a)$ in the left hand side, and put $x = a$ in what remains,

$$\frac{k}{()(a-b)}.$$

[†] See *A-level Mathematics*, also by F. G. J. Norton and published by HLT Publications.

MATHEMATICS REVISION WORKBOOK

This is A, the numerator of the fraction with $x - a$ as the denominator. Now cover up $(x - b)$, and put $x = b$ in what remains,

$$\frac{k}{(b-a)(\quad)}.$$

This is B, the numerator of the fraction with $x - b$ as denominator. Here, we cover up $(2x - 1)$ and put $x = \tfrac{1}{2}$ in $\dfrac{1}{x^2(\quad)}$, giving $C = 4$.

We cover up x^2, put $x = 0$ in what remains, $\dfrac{1}{(\quad)(2x-1)}$, and obtain $B = -1$, the numerator of the fraction with x^2 as denominator. Notice that we have the numerator when x^2 is the denominator, not the first fraction with x. If we only 'covered up' x, we should still have a factor x in the denominator, which would cause problems when we try to evaluate when $x = 0$.

•• Strictly $\displaystyle\int \frac{1}{x}\,dx = \ln|x|$. Here x and $2x - 1$ are both positive when $x = 2$ and when $x = 1$, so we can omit the modulus sign $|x|$. We have used $\ln x$ and $\ln(2x - 1)$; both equal $\ln 1 = 0$ when $x = 1$, and also $\ln 3 - \ln 2 = \ln(3/2)$.

(b) Here we have a product. It does not appear to be the derivative of a 'function of a function', so we consider integrating by parts. One factor $\ln(4x)$ simplifies on differentiating, giving $\dfrac{4}{4x}$, the other does not become much more complicated, so this looks hopeful.

Compare $\displaystyle\int x^3 (\ln 4x)\,dx$ with $\displaystyle\int u\,\frac{dv}{dx}\,dx.$

Take $\ln 4x$ as u, so $\dfrac{du}{dx} = \dfrac{4}{4x} = \dfrac{1}{x}$

and x^3 as $\dfrac{dv}{dx}$, so $v = \dfrac{x^4}{4}.$

Since the formula for integrating by parts is

$$\int u\,\frac{dv}{dx}\,dx = uv - \int v\,\frac{du}{dx}\,dx,$$

$$\int [\ln(4x)]\,x^3\,dx = [\ln 4x]\,\frac{x^4}{4} - \int \frac{x^4}{4}\cdot\frac{1}{x}\,dx$$

$$= \tfrac{1}{4}x^4 \ln 4x - \int \tfrac{1}{4}x^3\,dx$$

$$= \underline{\underline{\tfrac{1}{4}x^4 \ln 4x - \tfrac{1}{16}x^4 + C.}}$$

PARTIAL FRACTIONS

(c) We are given the substitution $x = 3\tan\theta$, so we write
$$\frac{dx}{d\theta} = 3\sec^2\theta$$

and
$$\int \frac{1}{9+x^2}\,dx = \int \frac{1}{9+x^2}\cdot\frac{dx}{d\theta}\,d\theta.$$
$$= \int \frac{1}{9+9\tan^2\theta}\,(3\sec^2\theta)\,d\theta.$$

But $1+\tan^2\theta = \sec^2\theta$, so the integral becomes
$$\int \frac{1}{9\sec^2\theta}\,(3\sec^2\theta)\,d\theta$$
$$= \int \tfrac{1}{3}\,d\theta$$
$$= \tfrac{1}{3}\theta.$$

Since $x = 3\tan\theta$, $\theta = \text{invtan}\,(\tfrac{1}{3}x)$, and
$$\underline{\underline{\int \frac{1}{9+x^2}\,dx = \tfrac{1}{3}\,\text{invtan}\,(\tfrac{1}{3}x) + C.}}$$

Note

- Since the integral is a function of x, to be integrated with respect to x, when we change it into a function of θ, that must be integrated with respect to θ, hence we need the factor $\dfrac{dx}{d\theta}$.

7.5 Denominator of the form $(x+a)(x+b)$; expansion in series

Question

Express $f(x) = \dfrac{4-x}{(1-x)(2-x)}$ in partial fractions.

Hence or otherwise, for $|x| < 1$, obtain the expansion of $f(x)$ in ascending powers of x up to and including the term in x^3, simplifying each coefficient.

(L)

Solution

Using the 'cover-up method', we can write down
$$\underline{\underline{\frac{4-x}{(1-x)(2-x)} = \frac{3}{1-x} - \frac{2}{2-x}.}}$$

49

MATHEMATICS REVISION WORKBOOK

To find the expansion of $f(x)$ in ascending powers of x, we notice that the second term's denominator begins $2-$ and we need $1- \ldots$ before we can use the binomial expansion with a negative index, so divide numerator and denominator of that term by 2,

$$f(x) = \frac{3}{1-x} - \frac{1}{1-\frac{1}{2}x}$$

$$= 3(1-x)^{-1} - (1-\tfrac{1}{2}x)^{-1}$$

$$= 3\left[1 + (-1)(-x) + \frac{(-1)(-2)}{1 \times 2}(-x)^2 + \frac{(-1)(-2)(-3)}{1 \times 2 \times 3}(-x)^3 \ldots\right]$$

$$- \left[1 + (-1)(-\tfrac{1}{2}x) + \frac{(-1)(-2)}{1 \times 2}(-\tfrac{1}{2}x)^2 + \frac{(-1)(-2)(-3)}{1 \times 2 \times 3}(-\tfrac{1}{2}x)^3 \ldots\right]$$

$$= 3[1 + x + x^2 + x^3 \ldots\ldots] - [1 + \tfrac{1}{2}x + \tfrac{1}{4}x^2 + \tfrac{1}{8}x^3 \ldots$$

$$= 2 + \frac{5}{2}x + \frac{11}{4}x^2 + \frac{23}{8}x^3 \ldots$$

Comments

• To find the coefficients by the 'cover up' method, we write

$$\frac{4-x}{(1-x)(2-x)} = \frac{A}{(1-x)} + \frac{B}{(2-x)} \quad \text{ie} \quad \frac{4-x}{(1-x)(2-x)} = \frac{A(2-x) + B(1-x)}{(1-x)(2-x)}.$$

Equating the numerators,

$$4 - x = A(2-x) + B(1-x).$$

Putting $x=1$, $3=A$, $A=3$. Putting $x=2$, $2=B(-1)$, $B=-2$, as before.

•• To check these coefficients, we substitute a value for x in $f(x)$ and in the expansion. Because of the fractions in the expansion, $x = 0.02$ will be the easiest value to use.

$$f(0.02) = \frac{3.98}{0.98 \times 1.98} = 2.05112348\ldots\ldots$$

and we can see this is the same value obtained from the expansion,

$$2 + \frac{5}{2}(0.02) + \frac{11}{4}(0.02)^2 + \frac{23}{8}(0.02)^3 \ldots = 2 + 0.05 + 0.0011 + 0.000\,023\ldots\ldots$$

8 BINOMIAL, LOGARITHMIC, EXPONENTIAL EXPANSIONS

8.1 Notes
8.2 Binomial expansion (positive index)
8.3 Binomial expansion (negative index); approximate value of $1/\sqrt{3}$
8.4 Binomial expansion (positive integer index)
8.5 Binomial expansion (negative integer index); logarithmic expansion
8.6 Exponential expansion

8.1 Notes

Factorial notation

$$n! = n(n-1)(n-2)(n-3)\ldots\ldots 3 \times 2 \times 1$$

so

$$\frac{n!}{m!} = n(n-1)(n-2)\ldots\ldots(m+1), \; n > m.$$

$\dfrac{n!}{r!(n-r)!}$ is written $\binom{n}{r}$ or ^nC_r or $_nC_r$.

Binomial expansion

$$(x+a)^n = x^n + nx^{n-1}a + \frac{n(n-1)}{2}x^{n-2}a^2 + \ldots\ldots\ldots \frac{n!}{r!(n-r)!}x^{n-r}a^r + \ldots\ldots a^n$$

for all values of x and a, when n is a positive integer.

$$(1+x)^n = 1 + nx + \frac{n(n-1)}{1 \times 2}x^2 + \frac{n(n-1)(n-2)}{1 \times 2 \times 3}x^3 + \ldots.$$

for all values of n if $-1 < x < 1$, and when $x = -1$ and/or $x = +1$ for some values of n.

Logarithmic series

$$\ln(1+x) = x - \tfrac{1}{2}x^2 + \tfrac{1}{3}x^3 - \tfrac{1}{4}x^4 \ldots + (-1)^{r-1}\frac{x}{r}\ldots\ldots$$

when $-1 < x \leq 1$.

MATHEMATICS REVISION WORKBOOK

Exponential series

$$e^x = 1 + x + \frac{x^2}{2!} + \frac{x^3}{3!} \ldots \frac{x^r}{r!} \ldots$$

for all values of x.

Trigonometric series

$$\sin x = x - \frac{x^3}{3!} + \frac{x^5}{5!} - \frac{x^7}{7!} \ldots$$

$$\cos x = 1 - \frac{x^2}{2!} + \frac{x^4}{4!} - \frac{x^6}{6!} \ldots$$

for all values of x. NB, x is measured in RADIANS.

8.2 Binomial expansion (positive index)

Questions

Write down and simplify the binomial expansion of $(1 + 2x)^{\frac{1}{2}}$ up to and including the x^2 term. Using the approximation $\cos x \approx 1 - \frac{1}{2}x^2$, show that, for small x,
$$(1 + 2x)^{\frac{1}{2}} \cos x \approx 1 + x - x^2.$$

(WJEC)

Solution

Using the binomial expansion,

$$(1 + 2x)^{\frac{1}{2}} = 1 + \tfrac{1}{2}(2x) + \frac{\tfrac{1}{2}(-\tfrac{1}{2})(2x)^2}{1 \times 2} \ldots$$

$$= 1 + x - \tfrac{1}{2}x^2 \text{ up to and including the term in } x^2.$$

Using the approximation given for $\cos x$,

$$\cos x (1 + 2x)^{\frac{1}{2}} \approx (1 - \tfrac{1}{2}x^2)(1 + x - \tfrac{1}{2}x^2)$$

$$= 1 + x - \tfrac{1}{2}x^2 - \tfrac{1}{2}x^2 \ldots$$

$$= \underline{1 + x - x^2}, \text{ up to and including the term in } x^2.$$

Check

In this question, we were given the final approximation that we had to obtain, so we do not need to check. If we had not been given the final approximation, we can check our result by giving x a small numerical value, eg $x = 0.01$, and remembering that in $\cos x$, x is in **radians**, so

$$(1.02)^{\frac{1}{2}} \cos 0.01 = 1.00989999 \ldots$$

and

$$1 + 0.01 - 0.01^2 = 1.0099,$$

so that our approximation seems likely to be correct.

BINOMIAL, LOGARITHMIC, EXPONENTIAL EXPANSIONS

8.3 Binomial expansion (negative index); approximate value of $1/\sqrt{3}$

Question

Obtain the expansion in ascending powers of x, up to and including the term in x^3, of

$$\frac{1+5x}{(1+2x)^{\frac{1}{2}}}, |x|<\tfrac{1}{2}.$$

By putting $x = 0.04$, deduce an approximate value of $\frac{1}{\sqrt{3}}$, giving your answer to three decimal places.

(AEB)

Solution

The binomial expansion of $(1+2x)^{-\frac{1}{2}}$ is

$$1 + (-\tfrac{1}{2})(2x) + \frac{(-\tfrac{1}{2})(-\tfrac{3}{2})}{1 \times 2}(2x)^2 + \frac{(-\tfrac{1}{2})(-\tfrac{3}{2})(-\tfrac{5}{2})}{1 \times 2 \times 3}(2x)^3$$

$$= 1 - x + \tfrac{3}{2}x^2 - \tfrac{5}{2}x^3 \ldots \ldots$$

so $\quad \dfrac{1+5x}{(1+2x)^{\frac{1}{2}}} = (1+5x)(1 - x + \tfrac{3}{2}x^2 - \tfrac{5}{2}x^3 \ldots \ldots$

$$= \underline{1 + 4x - \tfrac{7}{2}x^2 + 5x^3}, \text{ up to and including the term in } x^3.$$

Putting $x = 0.04$ in $\dfrac{1+5x}{(1+2x)^{\frac{1}{2}}}$, we have $\dfrac{1.2}{\sqrt{1.08}}$

$$= \frac{1.2}{0.6\sqrt{3}}, \text{ since } 108 = 3 \times 36, \sqrt{1.08} = \sqrt{3} \times \sqrt{0.36}$$

$$= \frac{2}{\sqrt{3}}.$$

Putting $x = 0.04$ in the expansion, we have

$$1 + 4(0.04) - 3.5(0.04)^2 - 5(0.04)^3$$

$$= 1.15472$$

so $\quad \dfrac{2}{\sqrt{3}} \approx 1.15472$, and $\dfrac{1}{\sqrt{3}} = 0.57736$

$$= \underline{\underline{0.577}}, \text{ to 3 decimal places.}$$

Comments

Our calculator gives $1/\sqrt{3} = 0.57735\ldots\ldots$, confirming our accuracy. We can also check the expansion of $(1+2x)^{-\frac{1}{2}}$ by putting $x = -0.01$ in $(1+2x)^{-\frac{1}{2}}$ and its expansion, $1.0101525\ldots\ldots$

8.4 Binomial expansion (positive integer index)

Question

Expand $(1 + ax)^8$ in ascending powers of x up to and including the term in x^2. The coefficients of x and x^2 in the expansion of $(1 + bx)(1 + ax)^8$ are 0 and -36 respectively. Find the values of a and b, given that $a > 0$ and $b < 0$.

(L)

Solution

We write down the first three terms of the binomial expansion of $(1 + ax)^8$ using

$$(1 + x)^n = 1 + nx + \frac{n(n-1)}{1 \times 2} x^2 \ldots\ldots\ldots$$

so

$$(1 + ax)^8 = 1 + 8ax + \frac{8 \times 7}{1 \times 2}(ax)^2 \ldots\ldots$$

$$= \underline{\underline{1 + 8ax + 28a^2x^2}} \ldots\ldots\ldots$$

Using this expansion,

$$(1 + bx)(1 + ax)^8 = (1 + bx)(1 + 8ax + 28a^2x^2 \ldots\ldots$$

$$= 1 + 8ax + 28a^2x^2 \ldots\ldots$$

$$+ bx + 8abx^2 \ldots\ldots$$

$$= 1 + (8a + b)x + (28a^2 + 8ab)x^2 \ldots$$

Since the coefficient of x is given to be 0,

$$8a + b = 0.$$

Since the coefficient of x^2 is given to be -36,

$$28a^2 + 8ab = -36.$$

We now have two simultaneous equations, the second of which we can simplify by dividing by 4 to give $7a^2 + 2ab = -9$.
Looking at these equations,

$$8a + b = 0 \quad (1)$$

and $\quad 7a^2 + 2ab = -9 \quad (2).$

Multiply (1) by $2a$, $\quad 16a^2 + 2ab = 0 \quad (3).$

Subtract (3) from (2),

$$-9a^2 = -9,$$

$\therefore \quad a^2 = 1$

$\therefore \quad$ $a = 1$, since we are given $a > 0$ so we can ignore the negative root $a = -1$. Substituting in (1), $8(1) + b = 0$ so $b = -8$ and we have $\quad \underline{\underline{a = 1, \ b = -8.}}$

BINOMIAL, LOGARITHMIC, EXPONENTIAL EXPANSIONS

8.5 Binomial expansion (negative integer index); logarithmic expansion

Question
Given that $y = (1 + 2x)^{-3}$ and that $|x| < \frac{1}{2}$, find and simplify a series in ascending powers of x up to and including the term in x^3 for

(a) y,
(b) $\ln y$.

(AEB)

Solution
(a) Using the binomial expansion

$$(1 + 2x)^{-3} = 1 + (-3)(2x) + \frac{(-3)(-4)(2x)^2}{1 \times 2} + \frac{(-3)(-4)(-5)(2x)^3}{1 \times 2 \times 3} \cdots$$

$$= \underline{1 - 6x + 24x^2 - 80x^3} \ldots$$

(b) $\ln y = \ln(1 + 2x)^{-3}$

$\qquad = (-3)\ln(1 + 2x)$

Now the expansion of $\ln(1 + x)$ is given in the formulae booklet,

so $\qquad \ln(1 + 2x) = 2x - \frac{(2x)^2}{2} + \frac{(2x)^3}{3} \cdots$

$\qquad\qquad\qquad = \underline{2x - 2x^2 + 8x^3/3} \ldots$

and $\ln y = -3\ln(1 + 2x) = \underline{-6x + 6x^2 - 8x^3}$, up to and including the term in x^3.

Comments
- It is always wise to check any expansion by giving x a suitable value. In this case, as the terms in the expansion have alternate signs, put $x = -0.01$.

$$(1 + 2(-0.01))^{-3} = 1.0624824\ldots,$$

and we identify the coefficients $1, 6, 24, 80 \ldots$ in the right hand side, the 80 being partly obscured by a digit from the next term. If our calculator has a large enough display, $x = 0.001$ shows

$$(1 + 2(-0.001))^{-3} = 1.006024080\ldots$$

8.6 Exponential expansion

Question
Find the expansion, in ascending powers of x up to and including x^3, of the function

$$f(x) = (1 - x)^2 e^{6x}.$$

55

Find also the coefficient of x^n in the expansion of f(x), and prove that the only values of n for which the coefficient of x^n is zero are 4 and 9.

(O & C)

Solution

By the binomial expansion,

Now $(1-x)^2 = 1 - 2x + x^2$,

and $e^{6x} = 1 + 6x + \dfrac{(6x)^2}{2!} + \dfrac{(6x)^3}{3!}$

$= 1 + 6x + 18x^2 + 36x^3 \ldots\ldots$

so $(1-x)^2 e^{6x} = (1 - 2x + x^2)(1 + 6x + 18x^2 + 36x^3 \ldots)$

$= 1 + 6x + 18x^2 + 36x^3 \ldots\ldots$
$\quad - 2x - 12x^2 - 36x^3 \ldots\ldots$
$\qquad\qquad x^2 + 6x^3 \ldots\ldots$

$= 1 + 4x + 7x^2 + 6x^3$, up to and including the term in x^3.

To find the term in x^n, we notice that the terms we need in the expansion of e^{6x} are those in x^{n-2}, x^{n-1} and x^n. Thus

$$(1-x)^2 = (1 - 2x + x^2)(\ldots\ldots\ldots \dfrac{(6x)^{n-2}}{(n-2)!} + \dfrac{(6x)^{n-1}}{(n-1)!} + \dfrac{(6x)^n}{n!} \ldots)$$

so the coefficient of x^n is

ie $\qquad \dfrac{6^n}{n!} - 2\dfrac{6^{n-1}}{(n-1)!} + \dfrac{6^{n-2}}{(n-2)!}$

$\qquad \dfrac{6^{n-2}}{n!}[36 - 2(6)(n) + n(n-1)]$

$= \dfrac{6^{n-2}}{n!}[n^2 - 13n + 36]$

$= \dfrac{6^{n-2}}{n!}(n-4)(n-9)$.

The only values of n for which this is zero are 4 and 9, so

$\underline{\underline{n = 4 \text{ and } n = 9}}$ are the only values of n for which the coefficient of x^n is zero.

Notes

It is easy to check the expansion of $(1-x)^2 e^{6x}$, by putting $x = 0.01$ in the expression and its expansion. When $x = 0.01$, $(1-x)^2 e^{6x} = 1.040705999\ldots$, and we see that the numerals 1, 4, 7, are the coefficients in the expansion. When $x = 0.01$, the value of the expansion is 1.040706, exactly the same apart from the small discrepancy between the last figures, as the coefficient of x^4 is zero and that of x^5 is negative.

9 SECTOR OF A CIRCLE

9.1   Notes
9.2   Perimeter and area of a circle
9.3   Perimeter and area; maximum value of the area

9.1   Notes

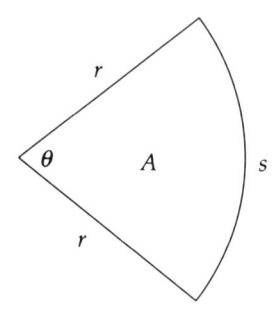

For a sector of a circle, radius r, angle θ measured in **RADIANS**,
  the arc length $s = r\theta$,
  the perimeter $p = r\theta + 2r$,
and the area $A = \tfrac{1}{2}r^2\theta$.

9.2   Perimeter and area of a circle

Question

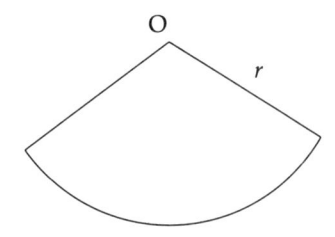

The diagram shows a sector of a circle, with the centre O and radius r. The length of the arc is equal to half the perimeter of the sector. Find the area of the sector in terms of r.   (C)

58

SECTOR OF A CIRCLE

Hint We shall need the formulae for the arc length s and the area A of a sector of a circle. If the angle at the centre is θ,
$$s = r\theta \quad \text{and} \quad A = \tfrac{1}{2}r^2\theta.$$

Solution

Draw our own diagram marking in the angle θ, the radius r, and the arc length $r\theta$.

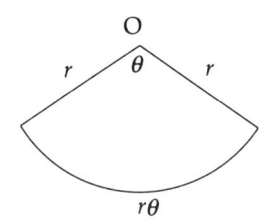

This reminds us that the perimeter is the whole of the boundary of the sector, so that the perimeter is $r\theta + r + r$, ie $r\theta + 2r$.

Since the arc length is half the perimeter,
$$r\theta = \tfrac{1}{2}(r\theta + 2r)$$
$$= \tfrac{1}{2}r\theta + r$$
$$\therefore \quad \tfrac{1}{2}r\theta = r,$$
$$\theta = 2. \quad \bullet$$

Since the area A is $\tfrac{1}{2}r^2\theta$,
$$\underline{\underline{A = r^2}}.$$

Comment

- Remember θ is in radians. 2 rad is about $106°$, which is reasonable. An angle of $2°$ would be suspiciously small.

9.3 Perimeter and area; maximum value of the area

Question

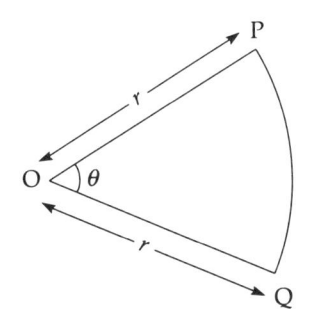

MATHEMATICS REVISION WORKBOOK

The figure shows a sector POQ of a circle radius r metres, the angle POQ being equal to θ radians. The perimeter of the sector is of length 3 metres and the area of the sector is A square metres.

(a) Show that
  $A = \frac{1}{2}r(3 - 2r)$.
(b) Show that, as r varies, the maximum value of A is 9/16.
(c) Find the corresponding value of θ.

(WJEC)

Hints

(i) Notice that we are given the perimeter of the sector is length 3 metres. That means that the arc length and the two radii total 3 metres,
ie
$$r\theta + 2r = 3.$$
(ii) We know the length of a circular arc is $r\theta$ and the area is $\frac{1}{2}r^2\theta$ so we can eliminate θ between $s = r\theta$ and $A = \frac{1}{2}r^2\theta$.

Solution

(a) Since $r\theta + 2r = 3,\quad r\theta = 3 - 2r$
  Since $\quad A = \frac{1}{2}r^2\theta,\quad A = \frac{1}{2}r(r\theta)$
  $\underline{= \frac{1}{2}r(3 - 2r).}$ •

(b) To find the maximum value of A, we write $A = \frac{3}{2}r - r^2$.

  Differentiating, $\dfrac{dA}{dr} = \dfrac{3}{2} - 2r$

  so $\dfrac{dA}{dr} = 0$ when $r = \frac{3}{4}$.

  When $r = \frac{3}{4}$ $A = \frac{1}{2}\left(\frac{3}{4}\right)\left(3 - 2 \times \frac{3}{4}\right)$
  $= 9/16.$

  Differentiating again, $\dfrac{d^2A}{dr^2} = -2$, which is negative, so that this value of A is a maximum,

  $\underline{\text{the maximum value of } A \text{ is } 9/16.}$ ••

(c) This maximum occurs when $r = \frac{3}{4}$. Since $r\theta = 3 - 2r$,

$$\theta = \frac{3 - 2\left(\frac{3}{4}\right)}{\frac{3}{4}} = 2,$$ •••

the maximum value of A occurs when $\underline{\theta = 2 \text{ radians.}}$

SECTOR OF A CIRCLE

Comments

- It is easier to equate the two expressions for $r\theta$, rather than just θ, ie $r\theta = 3 - 2r$ is simpler than $\theta = (3 - 2r)/r$.
- • We could find the maximum value by completing the square,

ie
$$A = \frac{3}{2}r - r^2$$
$$= \frac{9}{16} - \left(\frac{3}{4} - r\right)^2.$$

From this expression we can see that the greatest value of A occurs when the term in the brackets is 0, ie when $r = \frac{3}{4}$, and this greatest value is 9/16. This method is quicker than differentiating, but students too often prefer differentiating, perhaps because completing the square is found difficult. Here

$$A = \frac{3}{2}r - r^2$$
$$= \left(\frac{3}{4}\right)^2 - \left[\frac{9}{16} - \frac{3}{2}r + r^2\right], \quad \frac{3}{4} \text{ being 'half the coefficient of } r\text{', when the coefficient of } r^2 \text{ is 1.}$$

In the equation $A = \frac{3}{2}r - r^2$ the coefficient of r^2 is negative, so that we know the graph of A against r will be concave downwards, and that the only turning point will

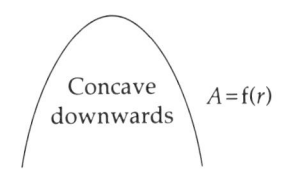

Concave downwards $A = f(r)$

be a maximum. This saves us the need of differentiating a second time, although that is very easy here.
- • • We may be more familiar with angles measured in degrees rather than radians. 2 radians is about 106°, which is reasonable.

10 SINE AND COSINE FORMULAE

10.1 Notes
10.2 Cosine formula
10.3 Cosine formula; deducing $\sin\theta$ from $\cos\theta$
10.4 Sine formula; small increment

10.1 Notes
Useful identities
$$\sin^2\theta + \cos^2\theta = 1,$$
$$\tan^2\theta + 1 = \sec^2\theta,$$
$$1 + \cot^2\theta = \operatorname{cosec}^2\theta.$$

The sine rule
$$\frac{a}{\sin A} = \frac{b}{\sin B} = \frac{c}{\sin C}.$$

The cosine rule
$$a^2 = b^2 + c^2 - 2bc\cos A,$$
whence
$$\cos A = \frac{b^2 + c^2 - a^2}{2bc}.$$

10.2 Cosine formula
Question

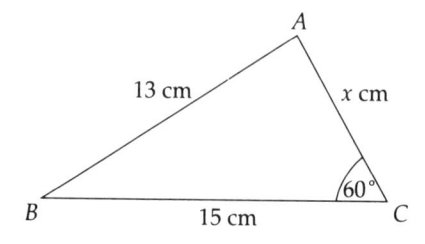

The figure shows a triangle ABC in which $AB = 13$ cm, $BC = 15$ cm and angle $ACB = 60°$. Given that $AC = x$ cm, calculate the possible values of x.

(L)

62

SINE AND COSINE FORMULAE

Hint Since angle C is the only angle we are given, we have to use the cosine formula with cos C, even although this gives a quadratic in x. We know we can solve the quadratic by the formula.

Solution

By the cosine formula,
$$13^2 = 15^2 + x^2 - 2 \times x \times 15 \cos 60°$$
$$169 = 225 + x^2 - 15x, \text{ using } \cos 60° = \tfrac{1}{2}$$
$$x^2 - 15x + 56 = 0$$
$$(x-7)(x-8) = 0, \text{ or by the formula,}$$
$$\underline{\underline{x = 7 \text{ or } 8.}}$$

10.3 Cosine formula; deducing sin θ from cos θ

Question

The lengths of the sides of a triangle are 4 cm, 5 cm and 6 cm. The size of the largest angle is θ.

(a) Calculate the value of $\cos \theta$.
(b) Hence show that $\sin \theta = \dfrac{a\sqrt{7}}{b}$, where a and b are integers that should be found.
(L)

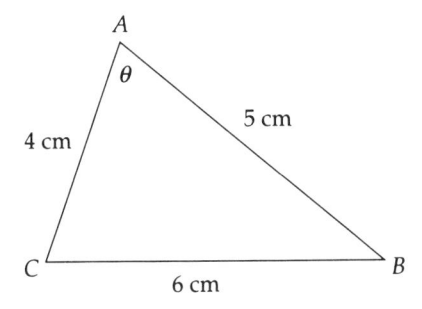

Hints Always draw a diagram.
Remember that in any triangle the largest side is opposite the largest angle.

Solution

(a) The cosine formula gives $\quad \cos A = \dfrac{b^2 + c^2 - a^2}{2bc}$

so $\quad \cos \theta = \dfrac{4^2 + 5^2 - 6^2}{40}$

$\underline{\underline{= \tfrac{1}{8}.}}$

63

(b) Since $\cos^2\theta + \sin^2\theta = 1$,
$$\sin^2\theta = 1 - \left(\tfrac{1}{8}\right)^2$$
$$= \frac{63}{64}$$
so $$\sin\theta = \frac{3\sqrt{7}}{8}$$

which is of the form required, with $\underline{\underline{a = 3, b = 8}}$.

Comments

- We should know whether our booklet of formulae gives both
$$a^2 = b^2 + c^2 - 2bc\cos A$$
and $$\cos A = \frac{b^2 + c^2 - a^2}{2bc}.$$

•• We required the exact value of $\sin\theta$ later, so we had to have the exact value of $\cos\theta$. Otherwise we should have gone straight from the first expression for $\cos\theta$ to the angle θ, by calculator.

10.4 Sine formula; small increment

Question

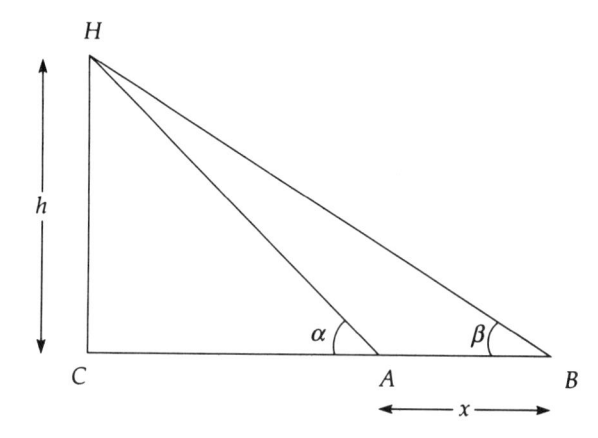

(i) The figure shows a helicopter H at a height h above a fixed point C at sea-level. The pilot observes two stationary ships A and B both due south of H. The angles of depression of A, B are α, β respectively. Show that the distance x between the ships is given by
$$x = \frac{h\sin(\alpha - \beta)}{\sin\alpha \sin\beta}.$$

SINE AND COSINE FORMULAE

(ii) Express h in terms of AC and α and show that

$$\frac{dh}{d\alpha} = 2h \operatorname{cosec} 2\alpha.$$

Hence write down an approximate expression for the increase $\delta\alpha$ in α when h is increased by an amount δh
The helicopter increases its height from 10 000 feet to 11 000 feet. Given that α was initially 35°, use your result to calculate, approximately, the consequent increase in α, giving your answer in degrees correct to 1 decimal place.

(WJEC)

Hint In (i), all the 'sines' in the expression for x suggest that we shall use the sine rule on triangle ABH, having found AH in terms of h first of all. Part (ii) deals with changes in h for small changes in α, when AC is constant, so we expect to differentiate the expression we find for h in terms of AC and α, and eliminate AC between the two equations.

Solution

(i) In triangle ACH, $\quad \dfrac{h}{AH} = \sin \alpha,$

so $\quad AH = h/\sin \alpha.$

Now apply the sine rule to triangle ABH, as we know the sides AH, AB and the angles opposite them.

$$\frac{AB}{\sin(\alpha - \beta)} = \frac{AH}{\sin \beta}$$

$$\underline{\underline{x = \frac{h \sin(\alpha - \beta)}{\sin \alpha \; \sin \beta}.}}$$

(ii) From triangle ACH,

$$\frac{h}{AC} = \tan \alpha \quad \text{so}$$

$$h = AC \tan \alpha. \tag{1}$$

Differentiating, keeping AC constant,

$$\frac{dh}{d\alpha} = AC \sec^2 \alpha \tag{2}$$

so eliminating AC between (1) and (2),

MATHEMATICS REVISION WORKBOOK

$$\frac{dh}{d\alpha} = \frac{h}{\tan \alpha} \sec^2 \alpha$$

$$= \frac{h}{\left(\frac{\sin \alpha}{\cos \alpha}\right)} \frac{1}{\cos^2 \alpha}$$

$$= \frac{h}{\sin \alpha \cos \alpha}$$

$$= \frac{2h}{2 \sin \alpha \cos \alpha}$$

$$= \frac{2h}{\sin 2\alpha}$$

$$\frac{dh}{d\alpha} = 2h \operatorname{cosec} 2\alpha.$$

Since $\quad \delta h \approx \dfrac{dh}{d\alpha} \delta\alpha$

$$\underline{\underline{\delta h = 2h \operatorname{cosec} 2\alpha \, \delta\alpha.}}$$

When $\alpha = 35°$, h increases from $10\,000$ to $11\,000$, so $\delta h = 1000$. Substituting, we have

$$1000 = 2 \times 10\,000 \operatorname{cosec} 70° \, \delta\alpha$$

$$\therefore \quad \delta\alpha = \frac{1000}{20\,000 \operatorname{cosec} 70°},$$

$$= 0.04698\ldots\ldots$$

This value is in radians, so the change in degrees is $0.04698\ldots \times 180 \div \pi$

$$\underline{\underline{= 2.7°, \text{ to 1 decimal place.}}}$$

Comments

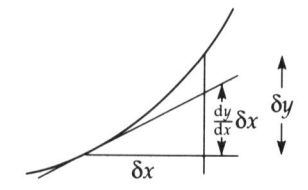

•

Since the gradient of the tangent at a point (x, y) is dy/dx, small increases δy in y are given approximately by

$$\delta y = \frac{dy}{dx} \delta x$$

as illustrated in the diagram.

SINE AND COSINE FORMULAE

•• We have to remember that when we differentiate trig functions the angle is almost invariably measured in radians. We do not need to change 35° into radians, as we are finding cosec of that angle, which is the same whatever the units in which the angle is measured, but we do have to realise that α is in radians, and change it into degrees to answer this question. We can check the reasonableness of this approximation by calculating AC when $h = 10\,000$ and $\alpha = 35°$, then calculating α for this value of AC and $h = 11\,000$. We find that $\alpha = 37.6°$, so the change in α is 2.6°, and our approximation was quite a good one.

11 SUMS AND PRODUCTS FORMULAE; SOLUTION OF TRIGONOMETRIC EQUATIONS

11.1 Notes
11.2 Sin $(A+B)$; expansion; error
11.3 Equations of the form $a \cos x + b \sin x = c$
11.4 Equations that can be reduced to quadratics
11.5 Equations that can be reduced to quadratics; use of sums and products formulae
11.6 Equations requiring use of sums and products formulae
11.7 Equations requiring use of $\sin 3\theta$ formula; general solution in degrees
11.8 General solutions in radians
11.9 Use of $\sin 3\theta$ to solve a cubic

11.1 Notes

Double angle formulae

$$\sin 2\theta = 2 \sin \theta \cos \theta,$$
$$\cos 2\theta = \cos^2 \theta - \sin^2 \theta$$
$$= 2\cos^2 \theta - 1$$
$$= 1 - 2\sin^2 \theta,$$
$$\tan 2\theta = \frac{2\tan \theta}{1 - \tan^2 \theta}.$$

SUMS AND PRODUCTS FORMULAE

Sums and products formulae

$$\sin(A + B) = \sin A \cos B + \cos A \sin B,$$
$$\sin(A - B) = \sin A \cos B - \cos A \sin B,$$
$$\cos(A + B) = \cos A \cos B - \sin A \sin B,$$
$$\cos(A - B) = \cos A \cos B + \sin A \sin B$$
$$\tan(A + B) = \frac{\tan A + \tan B}{1 - \tan A \tan B}$$
$$\tan(A - B) = \frac{\tan A - \tan B}{1 + \tan A \tan B}$$

$$\sin x + \sin y = 2 \sin \tfrac{1}{2}(x + y) \cos \tfrac{1}{2}(x - y)$$
$$\sin x - \sin y = 2 \cos \tfrac{1}{2}(x + y) \sin \tfrac{1}{2}(x - y)$$
$$\cos x + \cos y = 2 \cos \tfrac{1}{2}(x + y) \cos \tfrac{1}{2}(x - y)$$
$$\cos x - \cos y = -2 \sin \tfrac{1}{2}(x + y) \sin \tfrac{1}{2}(x - y).$$

Half-angle formulae

When $t = \tan \tfrac{1}{2} A$,

$$\sin A = \frac{2t}{1 + t^2}, \quad \cos A = \frac{1 - t^2}{1 + t^2}, \quad \tan A = \frac{2t}{1 - t^2}.$$

Auxiliary angle

$$a \sin \theta + b \cos \theta = R \sin(\theta + \alpha)$$

where $R = \sqrt{(a^2 + b^2)}$, and $\cos \alpha : \sin \alpha : 1 = a : b : R$
whence the greatest possible value of $a \sin \theta + b \cos \theta$ is $\sqrt{(a^2 + b^2)}$, the least possible value is $-\sqrt{(a^2 + b^2)}$.

Solution of equations, in degrees

If $\sin \theta = \sin \alpha$, $\theta = 360n° + \alpha$ or $(360n + 180)° - \alpha$:

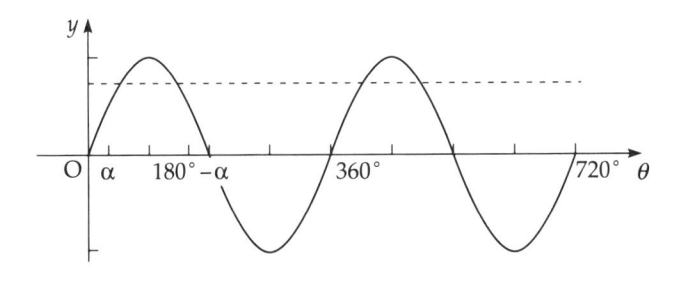

MATHEMATICS REVISION WORKBOOK

If $\cos\theta = \cos\alpha$, $\theta = 360n° + \alpha$ or $360n° - \alpha$:

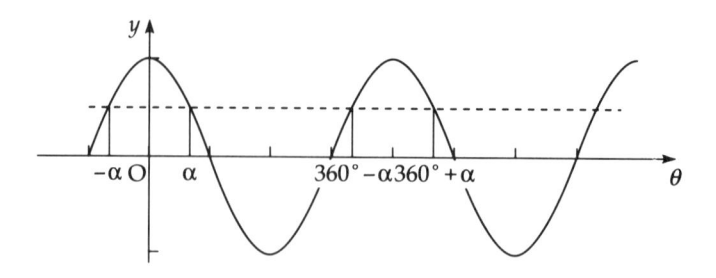

If $\tan\theta = \tan\alpha$, $\theta = 180n° + \alpha$.

Solution of equations, in radians

If $\sin\theta = \sin\alpha$, $\theta = 2n\pi + \alpha$ or $(2n+1)\pi - \alpha$
If $\cos\theta = \cos\alpha$, $\theta = 2n\pi + \alpha$ or $2n\pi - \alpha$
If $\tan\theta = \tan\alpha$, $\theta = n\pi + \alpha$

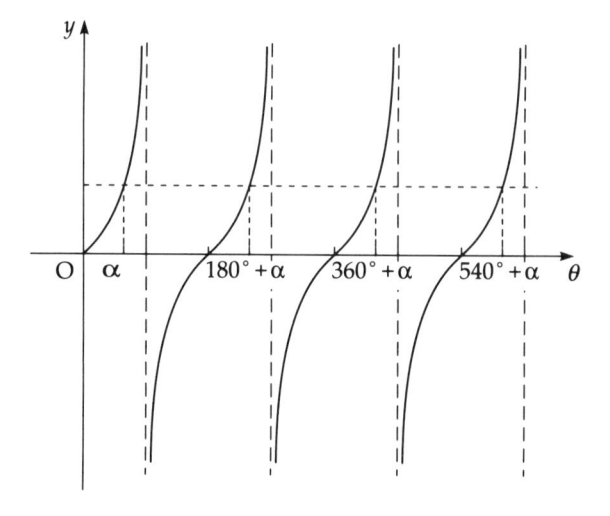

11.2 Sin $(A+B)$; expansion; error

Question

Show that, for a small angle t radians,

$$\sin(\pi/4 + t) \approx \frac{1}{\sqrt{2}}(1 + t - \tfrac{1}{2}t^2).$$

Use your calculator to estimate the error in using this approximation when $t = 0.1$, giving your answer to 5 decimal places. (L)

SUMS AND PRODUCTS FORMULAE

Hint   We shall use the formula for sin $(A + B)$, then the small-angle approximations
$$\sin x \approx x, \quad \cos x \approx 1 - \tfrac{1}{2}x^2.$$

Solution

Since $\sin(A + B) = \sin A \cos B + \cos A \sin B$
$$\sin(\pi/4 + t) = \sin(\pi/4) \cos t + \cos(\pi/4) \sin t$$
$$= \frac{1}{\sqrt{2}} \cos t + \frac{1}{\sqrt{2}} \sin t$$
$$= \frac{1}{\sqrt{2}}(1 - \tfrac{1}{2}t^2 + t)$$
$$= \underline{\underline{\frac{1}{\sqrt{2}}(1 + t - \tfrac{1}{2}t^2)}}.$$

Using our calculator,
$$\sin(\pi/4 + 0.1) = 0.774\ 167\ 078\ 5$$
and
$$\frac{1}{\sqrt{2}}(1 + 0.1 - \tfrac{1}{2}(0.1))^2 = 0.774\ 281\ 925$$

so the error in using this approximation is

0.000 114 8

ie $\underline{\underline{0.000\ 11}}$, to 5 dp.

Comment

Instead of finding the value of each of the expressions, it is easier and quicker to evaluate $\frac{1}{\sqrt{2}}(1 + 0.1 - \tfrac{1}{2}(0.1)^2) - \sin(\pi/4 + 0.1)$ on our calculator, obtaining $1.148\ 469\ldots \times 10^{-4}$, and deduce the error is 0.000 11, to 5 dp.

11.3 Equations of the form $a \cos x + b \sin x = c$

Question

(i)   Find the value of the acute angle α for which
$5 \cos x - 3 \sin x = \sqrt{34} \cos(x + \alpha)$ for all x.
Giving your answers correct to one decimal place,
(ii) solve the equation $5 \cos x - 3 \sin x = 4$ for $0° \leq x \leq 360°$,
(iii) solve the equation $5 \cos 2x - 3 \sin 2x = 4$ for $0° \leq x \leq 360°$.

(MEI)

MATHEMATICS REVISION WORKBOOK

Solution

(i) Using the expansion of $\cos(x + \alpha)$, we have
$$\sqrt{34}\cos(x + \alpha) = \sqrt{34}(\cos x \cos \alpha - \sin x \sin a)$$
$$\therefore \sqrt{34}(\cos x \cos \alpha - \sin x \sin \alpha) = 5\cos x - 3\sin x.$$
Since this is true for all values of x, the terms in $\cos x$ must be equal,
ie $\sqrt{34}\cos \alpha = 5$; the terms in $\sin x$ must be equal,
ie $\sqrt{34}\sin \alpha = 3$,
$$\tan \alpha = 3/5, \ \underline{\alpha = 30.96°, \text{ to 2 dp}}.$$

(ii) If $5\cos x - 3\sin x = 4$,
$$\sqrt{34}\cos(x + 30.96°) = 4,$$
$$\cos(x + 30.96°) = \frac{4}{\sqrt{34}}$$
$$x + 30.96° = 46.69 \text{ or } 313.31°$$
$$\underline{x = 15.7° \text{ or } 282.4°, \text{ to 1 dp}}.$$

(iii) Proceeding as in (ii), we have $\cos(2x + 30.96°) = \dfrac{4}{\sqrt{34}}$. As we have an equation in $2x$, we shall want the solutions for $2x$ between $0°$ and $720°$,
ie $\qquad 2x + 30.96° = 46.69° \text{ or } 313.31° \text{ or } 406.69° \text{ or } 673.31°$
$$\underline{x = 7.9° \text{ or } 141.2° \text{ or } 187.9° \text{ or } 321.2°}.$$

Comments

- As the answers are required correct to 1 decimal place, we must work with 2 dp. More accurate calculations show that
$$x = 15.7223° \ldots \text{ or } 282.3501° \ldots$$
so that our values given in (ii) and (iii) are correct to 1 dp, though the 282.4 is only just correct. Always work with at least one extra decimal place; when values can be stored in a calculator, so much the better.

11.4 Equations that can be reduced to quadratics

Question

Solve the equation
$$4\tan^2 x + 12\sec x + 1 = 0$$
giving all the solutions in degrees, correct to the nearest degree, in the interval $-180° < x < 180°$. (AEB)

Hint We should recall $\sec^2 x = 1 + \tan^2 x$, so that this becomes a quadratic in $\sec x$.

72

SUMS AND PRODUCTS FORMULAE

Solution

Since
$$4\tan^2 x + 12\sec x + 1 = 0$$
$$4(\sec^2 x - 1) + 12\sec x + 1 = 0,$$
$$4\sec^2 x + 12\sec x - 3 = 0,$$
$$\sec x = \frac{-12 \pm \sqrt{(12^2 - 4 \times 4 \times (-3))}}{8}$$
$$= \frac{-12 \pm \sqrt{192}}{8}.$$

Now $\sec x$ is $1/\cos x$, and $-1 \leq \cos x \leq 1$ for all real x, so that we can only find values of x if
$$\sec x \geq 1 \text{ or } \sec x \leq -1.$$

$(-12 + \sqrt{192}) \div 8 \approx 0.23$, so there are no real values of x for which $\sec x = (-12 + \sqrt{192}) \div 8$, and we need only look at the other root. If $\sec x = (-12 - \sqrt{192})/8$,

$x = 108°$, to the nearest degree, by calculator. The graph of $y = \sec x$ shows that there is another root, $x = -108°$, between $-180°$ and $180°$, so the solutions are
$$\underline{x = -108° \text{ or } 108°}.$$

Comments

We could multiply $4\tan^2 x + 12\sec x + 1 = 0$ by $\cos^2 x$, giving
$$4\sin^2 x + 12\cos x + \cos^2 x = 0,$$
$$4(1 - \cos^2 x) + 12\cos x + \cos^2 x = 0,$$
$$3\cos^2 x - 12\cos x - 4 = 0,$$
and then solve this quadratic. One root is greater than 1, so there will be no real values of x for which $\cos x > 1$; the other root gives the values $\pm 108°$.

11.5 Equations that can be reduced to quadratics; use of sums and products formulae

Question

(a) Solve the equation
$$3\sin^2 \theta = \cos \theta - 1$$
for all values of θ, $0 \leq \theta \leq 2\pi$.

(b) Express $\frac{1}{2}(\cos \theta + \cos 3\theta)$ as a product of two cosines. Hence or otherwise solve the equation
$$\cos \theta + \cos 3\theta = \cos 2\theta$$
for all values of θ, $0 \leq \theta \leq 2\pi$.

(O&C)

MATHEMATICS REVISION WORKBOOK

(a) Solution

Since $\cos^2\theta + \sin^2\theta = 1$ for all values of θ,
$$3\sin^2\theta = \cos\theta - 1 \Leftrightarrow 3(1 - \cos^2\theta) = \cos\theta - 1$$
ie
$$3\cos^2\theta + \cos\theta - 4 = 0$$
$$(3\cos\theta + 4)(\cos\theta - 1) = 0$$
$$\cos\theta = -4/3 \text{ or } \cos\theta = 1.$$

There are no real values of θ for which $\cos\theta = -4/3$, so the only real solutions come from $\cos\theta = 1$,

ie $\qquad\qquad\qquad\qquad\qquad \underline{\theta = 0 \text{ or } 2\pi.}$

Comment

- Since $1 - \cos^2\theta = (1 - \cos\theta)(1 + \cos\theta)$, we could have spotted that $\cos\theta - 1$ is a factor of both sides of this equation, so either $\cos\theta - 1 = 0$ or $-3(1+\cos\theta) = 1$, $\cos\theta = 1$ or $-4/3$ as before.

(b) Solution

From the formula in the booklet of formulae,
$$\tfrac{1}{2}(\cos\theta + \cos 3\theta) = \tfrac{1}{2}[2\cos 2\theta \cos\theta]$$
$$= \cos 2\theta \cos\theta.$$

If $\qquad\qquad \cos\theta + \cos 3\theta = \cos 2\theta,$
$\qquad\qquad\qquad \cos\theta \cos 2\theta = \cos 2\theta.$
Either $\qquad\quad \cos 2\theta = 0$ or $2\cos\theta = 1$, ie $\cos\theta = \tfrac{1}{2}.$
When $\qquad\quad \cos 2\theta = 0,\ 2\theta = \pi/2$ or $3\pi/2$ or $5\pi/2$ or $7\pi/2\ldots$
ie $\qquad\qquad \theta = \pi/4$ or $3\pi/4$ or $5\pi/4$ or $7\pi/4$,
keeping to the interval $0 \le \theta \le 2\pi$.
When $\cos\theta = \tfrac{1}{2},\ \theta = \pi/3$ or $5\pi/3$, in the same interval, so that, in the interval $0 \le \theta \le 2\pi$, the solutions are
$$\underline{\pi/4,\ 3\pi/4,\ 5\pi/4,\ 7\pi/4,\ \pi/3,\ 5\pi/3.}$$

Comment

It would probably be more useful to order these solutions
$$\pi/4,\ \pi/3,\ 3\pi/4,\ 5\pi/4,\ 5\pi/3 \text{ or } 7\pi/4$$
but it is easier to see where each comes from and that we have the complete set when listed as in the solution.
We can verify these solutions on our graphics calculator.

SUMS AND PRODUCTS FORMULAE

11.6 Equations requiring use of sums and products formulae

Question

Determine the value of x in degrees between $0°$ and $360°$ for which
$$2\sin x + \cos(x + 30°) = 0.$$
(AEB)

Solution

$$2\sin x + \cos(x + 30°) = 0$$
$$2\sin x + \cos x \cos 30° - \sin x \sin 30° = 0$$
$$\sin x (2 - \sin 30°) + \cos x \cos 30° = 0$$
$$\sin x = -\frac{\cos 30°}{2 - \sin 30°} \cos x$$
$$\tan x = \frac{-\cos 30°}{2 - \sin 30°} = -\frac{2\cos 30°}{3}.$$
By calculator, $\underline{x = 150° \text{ or } 330°}$.

Comments

- The exact value of $\cos 30°$ is $\frac{1}{2}\sqrt{3}$, so the exact value of $\tan x$ here is $-\frac{1}{\sqrt{3}}$, which we may recognise as $\tan 150°$. If we had had a value other than $30°$ in the original equation we probably would not have had an exact value for $\tan x$, so it was likely to be best to evaluate the expression by calculator, then use invtan (ie $\tan^{-1}$).

11.7 Equations requiring use of sin 3θ formula; general solution in degrees

Question

Express $\dfrac{\sin 3\theta}{\sin \theta}$ in terms of $\cos \theta$.

Hence show that if $\sin 3\theta = \lambda \sin 2\theta$, where λ is a constant, then either $\sin \theta = 0$ or $4\cos^2 \theta - 2\lambda \cos \theta - 1 = 0$.
Determine the general solution, in degrees, of the equation
$$\sin 3\theta = 3\sin 2\theta.$$
(AEB)

75

MATHEMATICS REVISION WORKBOOK

Solution

Since
$$\sin 3\theta = \sin(2\theta + \theta)$$
$$= \sin 2\theta \cos \theta + \cos 2\theta \sin \theta$$
$$= 2\sin\theta \cos^2\theta + (2\cos^2\theta - 1)\sin\theta$$
$$\frac{\sin 3\theta}{\sin \theta} = 2\cos^2\theta + 2\cos^2\theta - 1$$
$$= \underline{4\cos^2\theta - 1}.$$

•

If $\sin 3\theta = \lambda \sin 2\theta$,

$$\frac{\sin 3\theta}{\sin \theta} = \frac{\lambda 2 \sin\theta \cos\theta}{\sin \theta}, \text{ if } \sin\theta \neq 0$$

ie $4\cos^2\theta - 1 = 2\lambda \cos\theta$

ie $\underline{4\cos^2\theta - 2\lambda\cos\theta - 1 = 0}$, or $\underline{\sin\theta = 0}$.

To solve $\sin 3\theta = 3 \sin 2\theta$, put $\lambda = 3$.
Then either $\sin\theta = 0$ or $4\cos^2\theta - 6\cos\theta - 1 = 0$.

If $4\cos^2\theta - 6\cos\theta - 1 = 0$, $\cos\theta = \frac{6 \pm \sqrt{(52)}}{8}$.

••

As $\frac{6 + \sqrt{52}}{8} > 1$, the only real values of θ are obtained from the root $\frac{6 - \sqrt{52}}{8}$, which are 98.7° and 261.3°.
The solutions from $\theta = 0$ are $0°, 180°, 360°$, etc so that the general solutions of this equation are
$$\underline{\theta = 180n° \text{ or } (360n \pm 98.7)°}.$$

Comments

• Using $\sin 2\theta = 2\sin\theta\cos\theta$ and $\cos 2\theta = 2\cos^2\theta - 1$.
•• Using the formula to solve a quadratic equation.

11.8 General solutions in radians

Question

Find, in radians, the general solution of the equation
$$\cos\theta = \sin 2\theta.$$

(L)

Solution

Since $\sin 2\theta = 2\sin\theta\cos\theta$,
$\cos\theta = \sin 2\theta \Rightarrow \cos\theta = 2\sin\theta\cos\theta$

76

SUMS AND PRODUCTS FORMULAE

either $\cos\theta = 0$ or $1 = 2\sin\theta$, ie $\sin\theta = \frac{1}{2}$.
If $\cos\theta = 0$, $\theta = \pi/2$ or $3\pi/2$ or $5\pi/2 \dots$ etc.
If $\sin\theta = \frac{1}{2}$, $\theta = \pi/6$ or $5\pi/6$ or $13\pi/6$ or $17\pi/6$, etc.
so the general solution is
$$\theta = (2n+1)\pi/2 \text{ or } 2n\pi + \pi/6 \text{ or } (2n+1)\pi - \pi/6.$$

Notes

To find the general solution, or indeed all the solutions in a given interval, it often helps to use our graphics calculator to display the graphs. Thus the graph of $y = \cos\theta$ shows that $\cos\theta = 0$ at $\theta = \pi/2$ and thereafter at values increasing by π, ie $n\pi + \pi/2$, ie $(2n+1)\pi/2$. The graph of $y = \sin\theta$ shows that $\theta = \pi/6$ or $5\pi/6$, and thereafter at intervals of 2π, ie $2n\pi + \pi/6$ or $2n\pi + 5\pi/6$, which we can write as $(2n+1)\pi - \pi/6$.

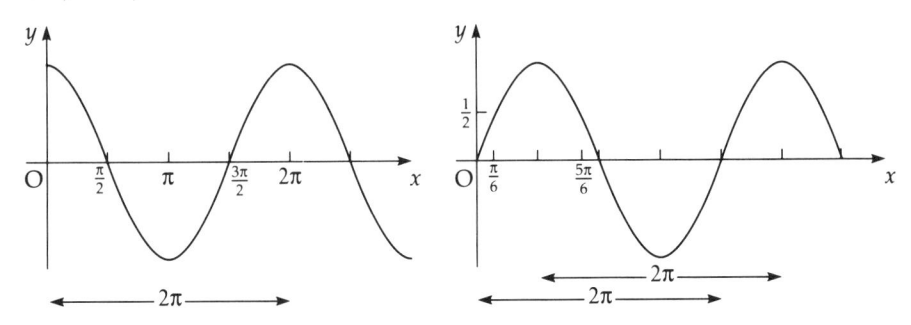

11.9 Use of sin 3θ to solve a cubic

Question

Prove the identity
$$\sin 3A = 3\sin A - 4\sin^3 A.$$
Hence show that $\sin 10°$ is a root of the equation
$$8x^3 - 6x + 1 = 0.$$

(AEB)

Solution

Since $\sin 3A = \sin(2A + A)$,
$\sin 3A = \sin 2A \cos A + \cos 2A \sin A$
$= (2\sin A \cos A)\cos A + (1 - 2\sin^2 A)\sin A$
$= 2\sin A \cos^2 A + \sin A - 2\sin^3 A$
$= 2\sin A(1 - \sin^2 A) + \sin A - 2\sin^3 A,$
$= 3\sin A - 4\sin^3 A.$

MATHEMATICS REVISION WORKBOOK

To show that $\sin 10°$ is a root of $8x^3 - 6x + 1 = 0$, we substitute $\sin 10°$ for x in the left hand side,

$$\begin{aligned}
8\sin^3 10° - 6\sin 10° + 1 &= 2(4\sin^3 10° - 3\sin 10°) + 1 \\
&= 2(-\sin 30°) + 1 \\
&= 2(-\tfrac{1}{2}) + 1 \\
&= 0
\end{aligned}$$

so $\sin 10°$ satisfies the equation $\underline{\underline{8x^3 - 6x + 1 = 0}}$.

Comments

- Using $\sin 2A = 2\sin A \cos A$ and $\cos 2A = 1 - 2\sin^2 A$.
- •• Putting $A = 10°$ in $3\sin A - 4\sin^3 A = \sin 3A$.

78

12 COORDINATE GEOMETRY OF THE STRAIGHT LINE AND CIRCLE

12.1 Notes

12.2 Straight line through two points; length of a line-segment

12.3 Straight line through two points; two perpendicular lines; gradient and ratio

12.4 Straight line through two points; perpendicular lines; point of intersection of two lines; ratio of lengths

12.5 Equation of a circle; gradient of a line; area of a triangle and of a sector; gradient of a tangent

12.1 Notes

In two dimensions,

if the point P_1 has coordinates (x_1, y_1), etc,

the **length** of the line-segment P_1P_2 is
$$\sqrt{[(x_1 - x_2)^2 + (y_1 - y_2)^2]},$$

the **midpoint** of P_1P_2 is
$$\left(\tfrac{1}{2}(x_1 + x_2), \tfrac{1}{2}(y_1 + y_2)\right),$$

the **gradient** of the line P_1P_2 is
$$\frac{y_2 - y_1}{x_2 - x_1},$$

the **equation** of the straight line through P_1 and P_2 is
$$\frac{y - y_1}{y_2 - y_1} = \frac{x - x_1}{x_2 - x_1}.$$

In three dimensions,
if the point P_3 has coordinates (x_1, y_1, z_1), etc.
the **length** of the line-segment P_1P_2 is
$$\sqrt{[(x_1 - x_2)^2 + (y_1 - y_2)^2 + (z_1 - z_2)^2]},$$

the **midpoint** of P_1P_2 is
$$\left(\tfrac{1}{2}(x_1 + x_2), \tfrac{1}{2}(y_1 + y_2), \tfrac{1}{2}(z_1 + z_2)\right).$$

MATHEMATICS REVISION WORKBOOK

Two straight lines with gradients m, m' are

$$\text{parallel if } m = m',$$
$$\text{perpendicular if } mm' = -1.$$

The equation of the circle centre (h, k), radius r is

$$(x - h)^2 + (y - k)^2 = r^2,$$

in particular, the equation of the circle, centre the origin, radius r is

$$x^2 + y^2 = r^2.$$

The circle with equation $x^2 + y^2 + 2gx + 2fy + c = 0$ has

$$\text{centre } (-g, -f) \text{ and radius } \sqrt{(g^2 + f^2 - c)}.$$

12.2 Straight line through two points; length of a line-segment

Question

The straight line through the points (1, 4) and (−3, −4) meets the coordinate axes in the points A and B. Find the area of a square having AB as one of its sides. (L)

Hint

Always draw a diagram when trying coordinate geometry questions.

Solution

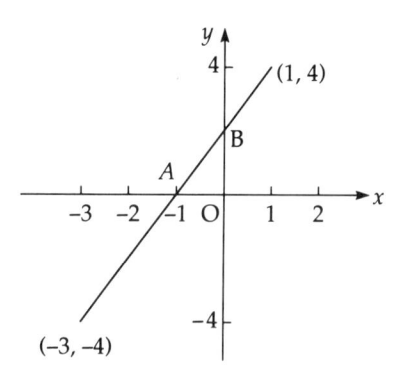

The equation of the straight line through points (x_1, y_1) and (x_2, y_2) is

$$\frac{y - y_1}{y_2 - y_1} = \frac{x - x_1}{x_2 - x_1}.$$

COORDINATE GEOMETRY OF THE STRAIGHT LINE AND CIRCLE

so the equation of the straight line through $(1, 4)$ and $(-3, -4)$ is

$$\frac{y-4}{-4-4} = \frac{x-1}{-3-1}$$

ie $-4(y-4) = -8(x-1)$

$$y - 4 = 2(x - 1)$$

$$y = 2x + 2.$$

This line meets the x-axis, ie $y = 0$ where $x = -1$, ie at $(-1, 0)$, and meets the y-axis, ie $x = 0$, where $y = 2$, ie at $(0, 2)$. Our diagram shows that these coordinates are reasonable, and also that $OA = 1$, $OB = 2$, so the length of AB is $\sqrt{5}$, and the area of the square on AB is

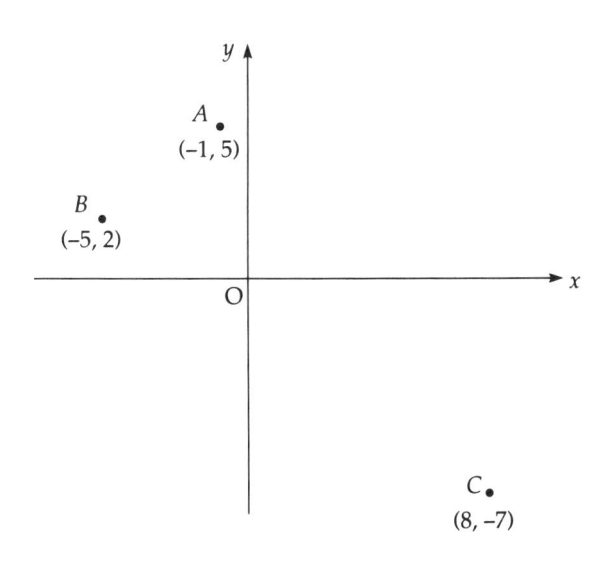

$$\underline{\underline{5}}.$$

12.3 Straight line through two points; two perpendicular lines; gradient and ratio

Question

The vertices of triangle ABC are the points $A(-1, 5)$, $B(-5, 2)$ and $C(8, -7)$.

(a) Find, in the form $px + qy + r = 0$, where p, q and r are integers, an equation of the line passing through B and C.
(b) Show by calculation that AB and AC are perpendicular.
  The point D lies on the line BA produced such that $3BA = BD$.
(c) Determine the coordinates of D.

(L)

Solution

(a)

Using the formula for the straight line through two points (page 79) the equation of BC is

$$\frac{y-2}{-7-2} = \frac{x+5}{8+5}$$

$$13(y-2) = -9(x+5),$$

ie $\qquad 13y - 26 = -9x - 45,$

$$9x + 13y + 19 = 0.$$

(b) The gradient of BA is

$$\frac{y_2 - y_1}{x_2 - x_1} = \frac{5-2}{-1-(-5)}$$

$$= \tfrac{3}{4}.$$

The gradient of AC is

$$\frac{-7-5}{8-(-1)}$$

$$= -\tfrac{4}{3}.$$

Two straight lines, gradients m_1, m_2 are perpendicular when $m_1 m_2 = -1$, and since $\tfrac{3}{4} \times -\tfrac{4}{3} = -1$, AB and AC are perpendicular.

(c)

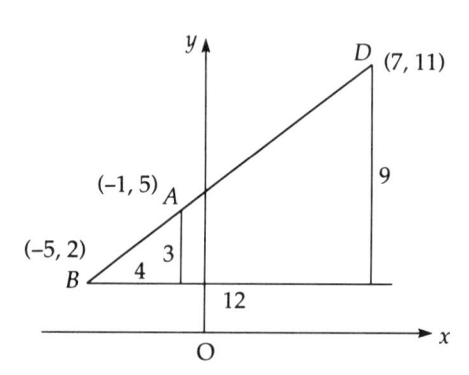

Since $3BA = BD$, the increase in x in going from B to D must be three times the increase in x going from B to A, ie the increase in x from B to D is 3×4, ie 12, so the x coordinate of D is $-5 + 12$, 7. Similarly, the increase in y is three times 3, ie 9 and the y coordinate of D is $2 + 9$, ie 11, so

$$D \text{ is}(7, 11).$$

COORDINATE GEOMETRY OF THE STRAIGHT LINE AND CIRCLE

12.4 Straight line through two points; perpendicular lines; point of intersection of two lines; ratio of lengths

Question

Points A, B and C have coordinates (1, 1), (2, 6) and (4, 2) respectively.
(a) Find an equation of the line through A and through the midpoint M of BC.
(b) Find an equation of the line through C which is perpendicular to BC.
(c) Solve these two equations to find the coordinates of the point P where these two lines meet.
(d) Find the ratio PA:AM, giving your answer in the form 1:k, where k is an integer.

(L)

Solution

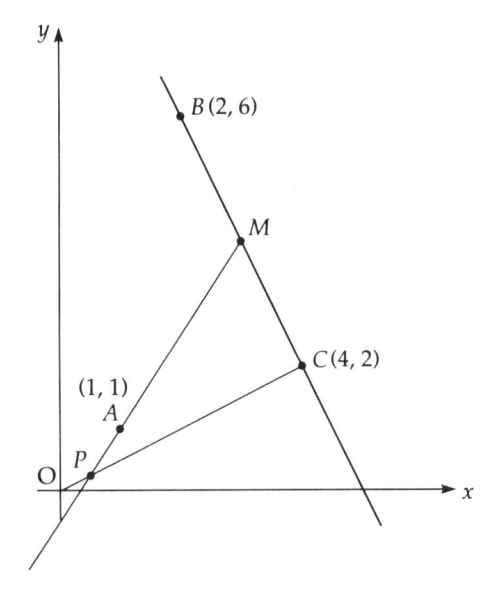

(a) The midpoint of the line joining (x_1, y_1) and (x_2, y_2) has coordinates

$$\left(\tfrac{1}{2}(x_1 + x_2), \tfrac{1}{2}(y_1 + y_2)\right),$$

so the midpoint of BC is

$$\left(\tfrac{1}{2}(2 + 4), \tfrac{1}{2}(6 + 2)\right), \text{ ie } (3, 4).$$

The equation of the line through (x_1, y_1), (x_2, y_2) is

$$y - y_1 = \frac{y_2 - y_1}{x_2 - x_1}(x - x_1)$$

83

MATHEMATICS REVISION WORKBOOK

so the line through (1, 1) and (3, 4) is $y - 1 = \frac{3}{2}(x - 1)$

ie
$$2(y - 1) = 3(x - 1),$$
$$\underline{\underline{2y = 3x - 1.}}$$

(b) The gradient of BC is -2, so the gradient of any line perpendicular to this is $\frac{1}{2}$, and the line through C perpendicular to BC is
$$y - 2 = \tfrac{1}{2}(x - 4),$$
$$\underline{\underline{2y = x.}}$$

(c) To find where $2y = 3x - 1$ and $2y = x$ intersect, solve them simultaneously, by substitution,

ie
$$x = 3x - 1,$$
$$x = \tfrac{1}{2}.$$

When $x = \tfrac{1}{2}$, since $2y = x$, $y = \tfrac{1}{4}$, and the coordinates of P are
$$\underline{\underline{(\tfrac{1}{2}, \tfrac{1}{4}).}}$$

(d)

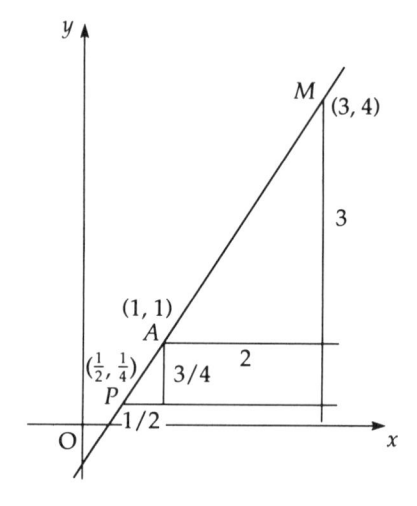

[Diagram: coordinate axes with points P($\tfrac{1}{2}$, $\tfrac{1}{4}$), A(1, 1), M(3, 4); similar triangles showing horizontal legs 3/4 and 2, vertical leg 3, and segment 1/2 from origin.]

From the diagram, $PA : AM = \tfrac{1}{2} : 2$
$$\underline{\underline{= 1 : 4.}}$$

Comment

- The similar triangles drawn in the diagram show that $PA : AM$ is equal to the ratio
  increase in x from P to A: increase in x from A to M

84

COORDINATE GEOMETRY OF THE STRAIGHT LINE AND CIRCLE

and is also equal to the ratio

increase in y from P to A: increase in y from A to M,

ie $PA : AM = \frac{1}{2} : 2$ and $PA : AM = \frac{3}{4} : 3$, both ratios being 1:4.

12.5 Equation of a circle; gradient of a line; area of a triangle and of a sector; gradient of a tangent

Question

The points $(5, 5)$ and $(-3, -1)$ are the ends of a diameter of the circle C with centre A. Write down the coordinates of A and show that the equation of C is

$$x^2 + y^2 - 2x - 4y - 20 = 0.$$

The line L with equation $y = 3x - 16$ meets C at the points P and Q. Show that P and Q satisfy the equation

$$x^2 - 11x + 30 = 0.$$

Hence find the coordinates of P and Q.
Write down the gradients of AP and AQ. Hence calculate the area of

(i) the triangle APQ,
(ii) the sector APQ.

Calculate the gradient of the tangent to C at P. Find the acute angle between L and this tangent.

(WJEC)

Solution

To draw a diagram we need the coordinates of A, the centre of the circle. Since A is the midpoint of $(5, 5)$ and $(-3, -1)$, the coordinates of A are

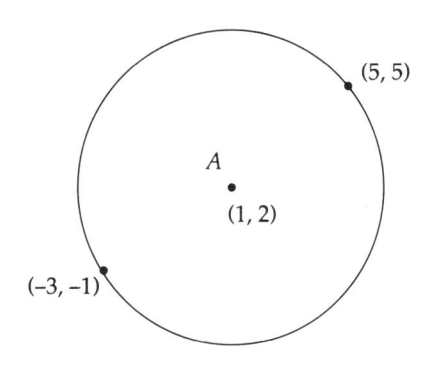

$$\tfrac{1}{2}(5 - 3), \tfrac{1}{2}(5 - 1) \text{ ie, } A \text{ is } (1, 2).$$

MATHEMATICS REVISION WORKBOOK

The distance of A from (5, 5) is $\sqrt{[(5-1)^2 + (5-2)^2]}$, ie $\sqrt{[4^2 + 3^2]}$, ie 5, so the equation of the circle is

$$(x-1)^2 + (y-2)^2 = 5^2,$$

ie
$$x^2 - 2x + 1 + y^2 - 4y + 4 = 25,$$
$$x^2 + y^2 - 2x - 4y - 20 = 0.$$

To find the coordinates of P and Q, solve the equation $y = 3x - 16$ simultaneously with that of the circle, $x^2 + y^2 - 2x - 4y - 20 = 0$, giving

$$x^2 + (3x - 16)^2 - 2x - 4(3x - 16) - 20 = 0,$$

ie
$$x^2 + 9x^2 - 96x + 256 - 2x - 12x + 64 - 20 = 0,$$
$$10x^2 - 110x + 300 = 0,$$
$$x^2 - 11x + 30 = 0,$$

so that P and Q satisfy the equation $x^2 - 11x + 30 = 0$. Factorising this expression,

$$(x - 5)(x - 6) = 0,$$
$$x = 5 \text{ or } x = 6,$$

When $x = 5, y = 3(5) - 16,$
$$= -1,$$
When $x = 6, \ y = 3(6) - 16 = 2$, so

the coordinates of P and Q are $(5, -1)$ and $(6, 2)$.

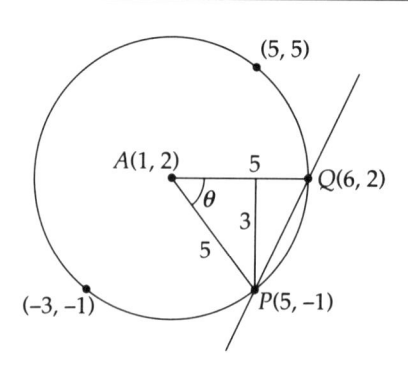

We can draw another diagram adding P and Q, to illustrate the final parts. From the diagram we can see that the gradient of AP is $\dfrac{-1-2}{5-1}$, ie $\dfrac{-3}{4}$ and the gradient of AQ is 0,

the gradient of AP is $-\dfrac{3}{4}$, the gradient of AO is 0.

COORDINATE GEOMETRY OF THE STRAIGHT LINE AND CIRCLE

(i) Taking AQ as the base of triangle APQ, area triangle APQ is $\frac{1}{2} \times 5 \times 3$, is 7.5.

(ii) The area of sector APQ is $\frac{1}{2}r^2\theta$, where $\tan\theta = \frac{3}{4}$ and θ is in radians,

so area sector $APQ = \frac{1}{2} \times 5^2 \times (\text{invtan } \frac{3}{4})$
$= 8.04$, to 2 dp.

Check

We can see that the area of the sector APQ is a little larger than the area of triangle APQ, so that the values 7.5 and 8.04 are reasonable.

Solution to final part

Since the gradient of AP is $-\frac{3}{4}$, the gradient of the tangent at P is $4/3$, as the tangent is perpendicular to the radius through the point of contact. To find the tangent of the angle θ between two lines gradients m_1, m_2, we use the formula

$$\tan\theta = \frac{m_1 - m_2}{1 + m_1 m_2}$$

where $m_1 = 4/3$, and $m_2 = 3$, the gradient of L,

so
$$\tan\theta = \frac{4/3 - 3}{1 + (4/3) \times 3}$$
$$= \frac{-5/3}{5}$$
$$= -\frac{1}{3}.$$

As this is negative, the acute angle will have tangent equal to $\frac{1}{3}$

Comment

The question does not say which of the points of intersection is P and which is Q. We have taken P to be the point with the smaller x coordinate, but the angles at P and Q are equal, so it does not matter whether we use AP or AQ.

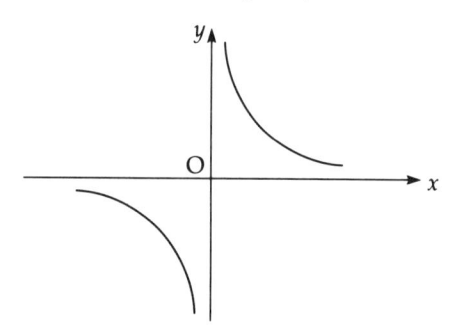

13 COORDINATE GEOMETRY USING PARAMETERS

13.1 Notes
13.2 The rectangular hyperbola $x = ct$, $y = c/t$
13.3 The curve $x = t^3$, $y = 3t^2$; finding dy/dx
13.4 The curve $x = t + e^t$, $y = t + e^{-t}$; finding dy/dx
13.5 The curve $x = t^2 - 2$, $y = t$; cartesian equation of a circle
13.6 Parameters in trigonometric form; finding dy/dx and d^2y/dx^2

13.1 Notes

The **parabola** $y^2 = 4ax$ can be written in parametric form
$$x = at^2, \quad y = 2at$$

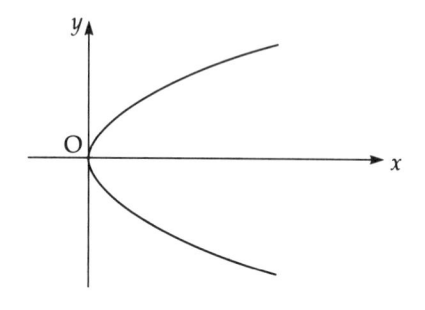

The **rectangular hyperbola** $xy = c^2$ can be written in parametric form
$$x = ct, \quad y = c/t$$

COORDINATE GEOMETRY USING PARAMETERS

The **semi-cubical parabola** $y^2 = x^3$ can be written in the form
$$x = t^2,\ y = t^3$$

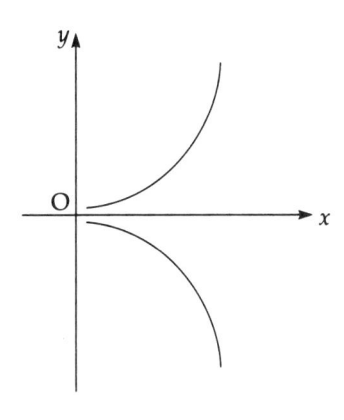

Derivatives

When x and y are expressed in terms of a parameter t, ie $x = x(t)$ and $y = y(t)$,
$$\frac{dy}{dx} = \frac{dy/dt}{dx/dt}.$$

13.2 The rectangular hyperbola $x = ct$, $y = c/t$

Question

(i) The points $P(cp, c/p)$, $Q(cq, c/q)$ lie on the rectangular hyperbola $xy = c^2$ (where c is a positive constant). Find the equation of the chord PQ.

(ii)

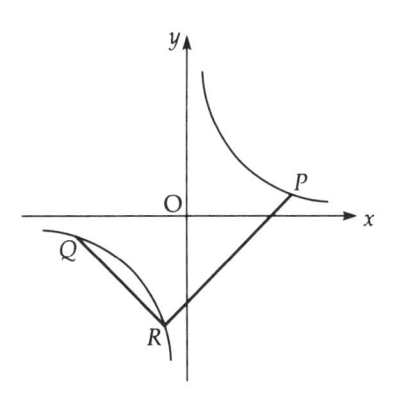

As shown in the figure, the points P and Q lie on different branches of the hyperbola. The point R $(cr, c/r)$ also lies on the hyperbola and is such

89

MATHEMATICS REVISION WORKBOOK

that PR, QR are perpendicular to each other. Prove that $r^2pq = -1$. Prove that PQ is perpendicular to the tangent at R.

(O & C)

Solution

(i) The equation of the straight line through two points (page 79) gives the equation of PQ as
$$\frac{y - c/p}{c/q - c/p} = \frac{x - cp}{cq - cp},$$
ie
$$\frac{pq(y - c/p)}{c(p - q)} = \frac{x - cp}{c(q - p)},$$
$$-pq(y - c/p) = (x - cp),$$
$$-pqy + cq = x - cp,$$
$$\underline{\underline{x + pqy = c(p + q).}}$$

(ii) From (i), the gradient of the chord PQ is $-1/pq$, so
  the gradient of PR is $-1/pr$,
  the gradient of QR is $-1/qr$.
Since PR and QR are perpendicular, $(-1/pr)(-1/qr) = -1$,
ie
$$r^2pq = -1.$$
The equation of the tangent at the point R $(cr, c/r)$ is
$$x + r^2y = 2cr$$
so the gradient of the tangent is $-1/r^2$. The gradient of PQ is $-1/pq$, and the product of the gradients is $(-1/r^2)(-1/pq)$, which is $1/r^2pq$. But we have already shown that this always has the value of -1, so
  <u>the tangent at R is always perpendicular to PQ.</u>

Comment

- We may be given the equation of the tangent in a list of formulae, or we may know it. If not, the equation can be deduced from the equation of the chord, by considering the tangent at P as the limit of the chord PQ when Q approaches P, so that we replace q by p in the equation of the chord. The chord through $(cp, c/p)$ and $(cq, c/q)$
$$x + pqy = c(p + q)$$
becomes the tangent at $(cp, c/p)$
$$x + p^2y = 2cp.$$

13.3 The curve $x = t^3$, $y = 3t^2$; finding dy/dx

Question

The curve C has parametric equations
$$x = t^3, \; y = 3t^2, \; t \geq 0.$$

COORDINATE GEOMETRY USING PARAMETERS

(a) Find $\frac{dy}{dx}$ in terms of t.

The points P and Q on C have parameters $t = 2$ and $t = 3$ respectively and O is the origin.

(b) Show that the tangent at P is parallel to the line OQ.

(L)

Solution

Since $x = t^3$, whereas y is only a multiple of t^2, x increases more rapidly than y, when $t \gg 1$, so the curve is as shown below:

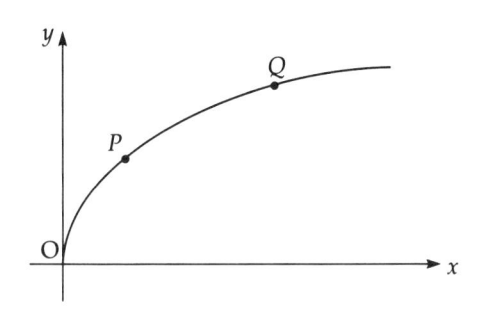

(a) Since $x = t^3$, $dx/dt = 3t^2$; since $y = 3t^2$, $dy/dt = 6t$,

so
$$\frac{dy}{dx} = \frac{dy/dt}{dx/dt}$$
$$= \frac{6t}{3t^2}$$
$$= \underline{\underline{2/t}}.$$

(b) At P, $t = 2$ so the gradient of the tangent at P is $2/2$, ie 1.
At Q, $t = 3$, so Q is $(27, 27)$, and the gradient of OQ is $27/27$, ie 1.
Since the gradients are equal, <u>the tangent at P is parallel to the line OQ</u>.

13.4 The curve $x = t + e^t$, $y = t + e^{-t}$; finding dy/dx

Question

The parametric equations of a curve C are
$x = t + e^t$, $y = t + e^{-t}$.

Find $\frac{dy}{dx}$ in terms of t, and hence find the coordinates of the stationary point of C.

(C)

Solution

Since $x = t + e^t$, $\dfrac{dx}{dt} = 1 + e^t$.

Since $y = t + e^{-t}$, $\dfrac{dy}{dt} = 1 - e^{-t}$

$$y = \frac{dy/dt}{dx/dt} = \frac{1 - e^{-t}}{1 + e^t}.$$

At the stationary point, $dy/dx = 0$, ie $1 - e^{-t} = 0$, $e^t = 1$, $t = 0$ so the coordinates of the stationary point are

$$\underline{\underline{(1,\ 1)}}.$$

This can be confirmed by a graphics calculator.

13.5 The curve $x = t^2 - 2$, $y = t$; cartesian equation of a circle

Question

Sketch the curve C defined parametrically by

$$x = t^2 - 2;\ y = t.$$

Write down the cartesian equation of the circle with centre the origin and radius r. Show that this circle meets the curve C at points whose parameter t satisfies the equation

$$t^4 - 3t^2 + 4 - r^2 = 0.$$

(a) In the case $r = 2\sqrt{2}$, find the coordinates of the two points of intersection of the curve and the circle.

(b) Find the range of values of r for which the curve and the circle have exactly two points in common.

(AEB)

Solution

If we have a graphics calculator, that will show us the form of the curve. Otherwise, we notice that as t increases from $-\infty$ to $+\infty$, y takes all values from $-\infty$ to $+\infty$. When t is large and negative, though, x is large but positive. As t increases from $-\infty$ to 0, x decreases from large and positive to the value -2, then x increases again, taking as large a value as we wish. The shape of the curve is below:

COORDINATE GEOMETRY USING PARAMETERS

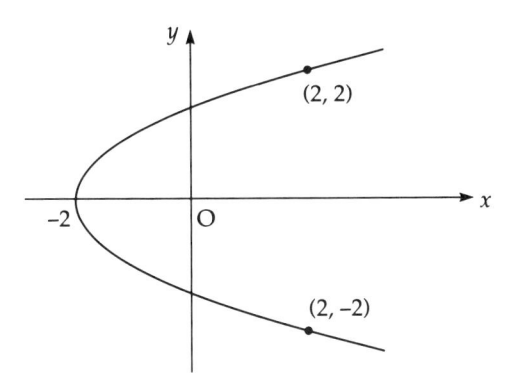

The equation of a circle, centre the origin, radius r is
$$x^2 + y^2 = r^2$$
and the points at which this meets $x = t^2 - 2$, $y = t$ are given by
$$(t^2 - 2)^2 + t^2 = r^2$$
ie
$$t^4 - 4t^2 + 4 + t^2 = r^2$$
$$t^4 - 3t^2 + 4 - r^2 = 0. \qquad (1)$$

(a) When $r = 2\sqrt{2}$, this equation becomes
$$t^4 - 3t^2 - 4 = 0,$$
which is a quadratic in t^2 and can be factorised
$$(t^2 - 4)(t^2 + 1) = 0,$$
so
$$t^2 = 4 \text{ or } -1,$$
ie
$$t = +2 \text{ or } -2,$$
and the points of intersection are $(2, 2)$ and $(2, -2)$.

(b) The curve and the circle have exactly two points in common if equation (1) has exactly two real values of t, ie the quadratic
$$z^2 - 3z + 4 - r^2 = 0$$
(writing z instead of t^2), must have two real roots, of which one is positive and one is negative. Any quadratic
$$z^2 - 3z + k = 0$$
will have one positive and one negative root only if k is negative (since the product of one positive root and one negative root is always negative), so that
$$z^2 - 3z + 4 - r^2 = 0$$
will have one positive and one negative root if and only if $4 - r^2$ is negative, ie
$$r^2 > 4,$$
$$\underline{\underline{r > 2.}}$$

Check

Sketching the curve C and circles of different radii illustrates the reasonableness of this result.

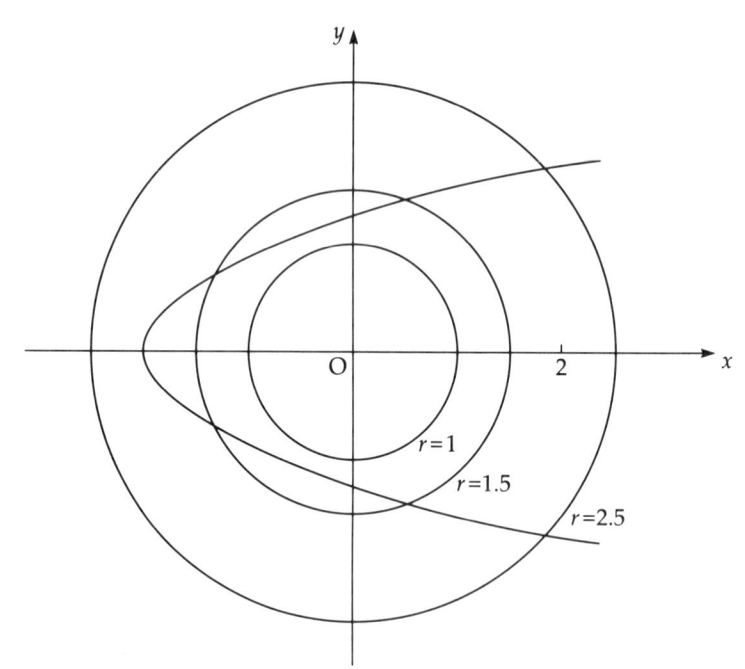

13.6 Parameters in trigonometric form; finding dy/dx and d²y/dx²

Question

A curve is defined for $-\dfrac{\pi}{6} < \theta < \dfrac{\pi}{6}$ by the parametric equations

$$x = \sin 4\theta + 2\sin 2\theta, \quad y = \cos 4\theta - 2\cos 2\theta.$$

Prove that $\dfrac{dy}{dx} = -\tan\theta$.

Find $\dfrac{d}{d\theta}\left(\dfrac{dy}{dx}\right)$ and show that $\dfrac{d^2y}{dx^2} = -\dfrac{1}{8}$ when $\theta = 0$.

(AEB)

Hint We find $\dfrac{dy}{dx}$ by finding $\dfrac{dx}{d\theta}$ and $\dfrac{dy}{d\theta}$ and using

$$\dfrac{dy}{dx} = \dfrac{dy}{d\theta} \div \dfrac{dx}{d\theta}$$

COORDINATE GEOMETRY USING PARAMETERS

BUT $\dfrac{d^2y}{dx^2} = \dfrac{d}{dx}\left(\dfrac{dy}{dx}\right) = \dfrac{d}{d\theta}\left(\dfrac{dy}{dx}\right) \cdot \dfrac{d\theta}{dx}$

$\qquad = \dfrac{d}{d\theta}\left(\dfrac{dy}{dx}\right) \div \dfrac{dx}{d\theta}.$

Solution

Since $\quad x = \sin 4\theta + 2\sin 2\theta,$

$\dfrac{dx}{d\theta} = 4\cos 4\theta + 4\cos 2\theta,$

$\qquad = 4\,[\cos 4\theta + \cos 2\theta]$
$\qquad = 4\,[2\cos 3\theta \cos \theta]$
$\qquad = 8\cos 3\theta \cos \theta.$

•

Similarly, $\quad y = \cos 4\theta - 2\cos 2\theta$

so $\qquad \dfrac{dy}{dx} = -4\sin 4\theta + 4\sin 2\theta$

$\qquad = -4\,[\sin 4\theta - \sin 2\theta]$
$\qquad = -4\,[2\cos 3\theta \sin \theta]$
$\qquad = -8\cos 3\theta \sin \theta.$

••

Thus $\quad \dfrac{dy}{dx} = \dfrac{\frac{dy}{d\theta}}{\frac{dx}{d\theta}} = \dfrac{-8\cos 3\theta \sin \theta}{8\cos 3\theta \cos \theta}$

$\qquad \underline{\underline{= -\tan \theta.}}$

Since $\qquad \dfrac{dy}{dx} = -\tan \theta$

$\dfrac{d}{d\theta}\left(\dfrac{dy}{dx}\right) = -\sec^2 \theta$

so $\qquad \dfrac{d^2y}{dx^2} = \dfrac{d}{d\theta}\left(\dfrac{dy}{dx}\right) \Big/ \dfrac{dx}{d\theta}$

$\qquad = \dfrac{-\sec^2 \theta}{8\cos 3\theta \cos \theta}$

$\qquad \underline{\underline{= -\tfrac{1}{8}}}$ when $\theta = 0$, since $\sec 0 = \cos 0 = 1.$

MATHEMATICS REVISION WORKBOOK

Comments

- Using 'sums and products' formulae,
$$\cos A + \cos B = 2\cos \tfrac{1}{2}(A+B)\cos \tfrac{1}{2}(A-B),$$
and
- - $$\sin A - \sin B = 2\sin \tfrac{1}{2}(A-B)\cos \tfrac{1}{2}(A+B).$$

14 DIFFERENTIATION

14.1 Notes
14.2 Derivative of ln f(x); derivative of a quotient
14.3 Derivative of a function of a function; derivative of a product; dy/dx when y is an implicit function of x
14.4 Derivative of a polynomial; point of inflexion
14.5 Derivative of an exponential; second derivative
14.6 Equation of a tangent to a curve; derivative of a quotient

14.1 Notes

Derivatives

Function f(x) or y	Derivative f'(x) or $\dfrac{dy}{dx}$
x^n	nx^{n-1}
$(ax+b)^n$	$na(ax+b)^{n-1}$
$\sin(ax+b)$	$a\cos(ax+b)$
$\cos(ax+b)$	$-a\sin(ax+b)$
$\tan(ax+b)$	$a\sec^2(ax+b)$
$\ln(ax+b)$	$\dfrac{a}{ax+b}$
e^{ax}	ae^{ax}

Function of a function

$$\frac{d}{dx} F[f(x)] = F'(f(x)) \frac{df}{dx}, \quad \text{eg} \quad \frac{d}{dx}(\sin 3x) = 3\cos 3x.$$

Product

$$\frac{d}{dx}(uv) = u\frac{dv}{dx} + v\frac{du}{dx}.$$

Quotient

$$\frac{d}{dx}\left(\frac{u}{v}\right) = \frac{v\frac{du}{dx} - u\frac{dv}{dx}}{v^2}.$$

Logarithmic function

$$\frac{d}{dx}\ln f(x) = \frac{f'(x)}{f(x)}, \quad \text{eg} \quad \frac{d}{dx}(\ln(x+\sin x)) = \frac{1+\cos x}{x+\sin x}.$$

MATHEMATICS REVISION WORKBOOK

Exponential function

$$\frac{d}{dx} e^{f(x)} = f'(x) e^{f(x)}, \qquad eg \quad \frac{d}{dx} (e^{\sin x}) = \cos x \, e^{\sin x}.$$

Stationary values

All points for which $f'(x) = 0$ are called stationary points, and the corresponding values of x are called **stationary values**. Thus maxima, minima and points of inflexion are stationary points. **Maxima** and **minima** (but not **points of inflexion**) are sometimes called turning points.

At a maximum, $f'(x) = 0$ and $f''(x)$ is negative as $f'(x)$ is decreasing.
At a minimum, $f'(x) = 0$ and $f''(x)$ is positive as $f'(x)$ is increasing.
At a point of inflexion, $f''(x) = 0$ and changes sign; $f'(x)$ may or may not equal zero.
'Near' a maximum, $f'(x)$ is positive, zero, then negative.
'Near' a minimum, $f'(x)$ is negative, zero, then positive.

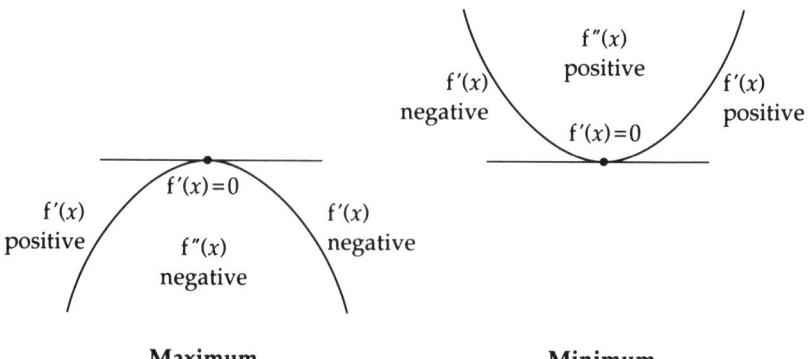

Maximum

Minimum

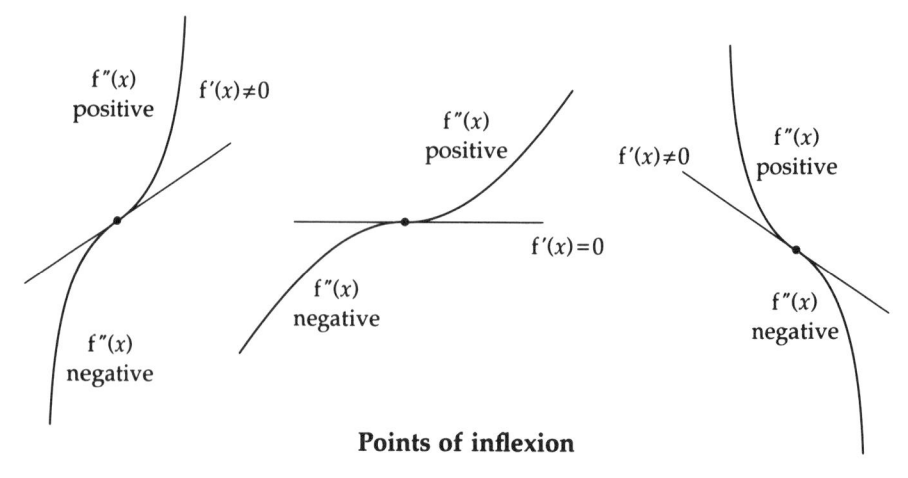

Points of inflexion

DIFFERENTIATION

14.2 Derivative of ln f(x); derivative of a quotient

Question

Differentiate with respect to x, simplifying your answers,

(i) $\ln(\sin^2 x)$,

(ii) $\dfrac{e^{2x}}{1+e^x}$.

(C)

Hint Since $\ln(a^2) = 2 \ln a$, we shall write $\ln(\sin^2 x)$ as $2 \ln(\sin x)$.
Remember that $\dfrac{d}{dx} \ln [f(x)] = \dfrac{f'(x)}{f(x)}$.

Solution

(i) $\dfrac{d}{dx} [\ln (\sin^2 x)] = \dfrac{d}{dx} [2 \ln(\sin x)]$

$= \dfrac{2 \cos x}{\sin x}$

$= 2 \cot x.$

(ii) Using the formula for the derivative of a quotient,

$$\dfrac{d}{dx}\left(\dfrac{u}{v}\right) = \dfrac{v \dfrac{du}{dx} - u \dfrac{dv}{dx}}{v^2},$$

$$\dfrac{d}{dx}\left(\dfrac{e^{2x}}{1+e^x}\right) = \dfrac{(1+e^x)2e^{2x} - e^{2x}(e^x)}{(1+e^x)^2}$$

$$= \dfrac{e^{2x}(2+e^x)}{(1+e^x)^2}.$$

Comment

- An alternative method depends on recognising that e^{2x} is $(e^x)^2$.

Consider $\dfrac{z^2}{z+1} = \dfrac{z^2 - 1 + 1}{z+1}$

$= \dfrac{(z-1)(z+1) + 1}{z+1}$

$= z - 1 + \dfrac{1}{z+1}$

so that $\dfrac{e^{2x}}{e^x + 1} = e^x - 1 + \dfrac{1}{e^x + 1}$

$= e^x - 1 + (e^x + 1)^{-1}.$

Differentiating with respect to x, $e^x - e^x(e^x + 1)^{-2}$
which we can write as $e^x - \dfrac{e^x}{(e^x + 1)^2}$.
We can rearrange this to show that it gives the same expression as the one we obtained by direct differentiation. This method of division is often useful when integrating.

14.3 Derivative of a function of a function; derivative of a product; dy/dx when y is an implicit function of x

Question

(a) Differentiate with respect to x
  (i) $\sqrt{(1 + 3x^3)}$, (ii) $x [\exp(x^2)]$, (iii) $\sin x \operatorname{cosec} 2x$.
(b) Given that $x^3 + y^3 + 2xy - x = 1$, find $\dfrac{dy}{dx}$ when $x = 2$, $y = -1$.

(O & C)

Solution

(a) (i) This is a 'function of a function', so write
$$f(x) = (1 + 3x^3)^{\frac{1}{2}}$$
Then $f'(x) = \frac{1}{2}(1 + 3x^3)^{-\frac{1}{2}} (9x^2)$
$$= \underline{\dfrac{9x^2}{2(1 + 3x^3)^{\frac{1}{2}}}}.$$

(ii) Here we have a product, so

if $f(x) = x [\exp (x^2)]$,
$f'(x) = [\exp (x^2)] + x (2x) \exp (x^2)$
$= \underline{(1 + 2x^2) [\exp (x^2)]}.$

(iii) As 'function of a function' and products have already been tested, we expect something different. We could do this as a product, $\sin x \operatorname{cosec} 2x$, or as a quotient, $\sin x \div \sin 2x$, but the easiest way is to realise that $\operatorname{cosec} 2x = 1 \div \sin 2x$, and $\sin 2x = 2 \sin x \cos x$, so that

$$\sin x \operatorname{cosec} 2x = \dfrac{\sin x}{\sin 2x} = \dfrac{\sin x}{2 \sin x \cos x} = \dfrac{1}{2 \cos x}$$
$$= \tfrac{1}{2} (\cos x)^{-1},$$

so $d (\sin x \operatorname{cosec} 2x)/dx = \tfrac{1}{2}(-1)(\cos x)^{-2}(- \sin x)$
$$= \underline{\dfrac{\sin x}{2 \cos^2 x}}.$$

DIFFERENTIATION

Notes
(i) Don't forget the derivative of the 'inside'. Here, $d(1 + 3x^3)/dx = 9x^2$, so we have a factor $9x^2$.
(ii) The notation $\exp(f(x))$ is often used to simplify printing where $e^{f(x)}$ would look awkward or be difficult to print. Thus we may have $e^{\sin x}$ printed as $\exp(\sin x)$, e^{-x^2} as $\exp(-x^2)$, etc. The square brackets [ ] are strictly not necessary, but are used to make the expressions perfectly clear.
(iii) As $(\cos x)^{-1} = \sec x$, we could have written
$$d(\sin x \operatorname{cosec} 2x)/dx = d\left(\tfrac{1}{2}\sec x\right)/dx$$
$$= \tfrac{1}{2}\sec x \tan x,$$ a simpler form perhaps than the one we gave.

(b) Here we have an implicit function, so we proceed by differentiating with respect to x,
$$3x^2 + 3y^2 \frac{dy}{dx} + 2\left(y + x\frac{dy}{dx}\right) - 1 = 0. \qquad \bullet$$

As we only want dy/dx for particular values of x and y, substitute $x = 2$ and $y = -1$ in this equation,
$$3(2)^2 + 3(-1)^2 \frac{dy}{dx} + 2\left(-1 + 2\frac{dy}{dx}\right) - 1 = 0,$$
$$\frac{dy}{dx} = -\frac{9}{7}.$$

Notes

If we had required dy/dx as a function of x and y, we should have had to collect terms in $\bullet$,
$$(3y^2 + 2x)\frac{dy}{dx} = 1 - 3x^2 - 2y,$$
$$\frac{dy}{dx} = \frac{1 - 3x^2 - 2y}{(3y^2 + 2x)}.$$

14.4 Derivative of a polynomial; point of inflexion

Question

The function f is defined for all real x by
$$f(x) = x^3 + 3x^2 + 4x.$$
Find $f'(x)$, the derivative of $f(x)$. By writing $f'(x)$ in the form
$$a(x + 1)^2 + b,$$
where a and b are constants, prove that $f(x)$ is positive when $x > 0$. Sketch the graph of
$$y = f(x),$$
marking the point of inflexion. (MEI)

Solution

Since $f(x) = x^3 + 3x^2 + 4x$,
$f'(x) = 3x^2 + 6x + 4$.

Now $3x^2 + 6x + 4 = 3x^2 + 6x + 3 + 1$
$= 3(x^2 + 2x + 1) + 1$,
$= 3(x+1)^2 + 1$,

which is in the form $a(x+1)^2 + b$.

As $(x+1)^2$ is a perfect square, $(x+1)^2 \geq 0$,
$\therefore \qquad (x+1)^2 + 1 > 0$,

so $f'(x)$ is never negative. But $f(0) = 0$, so when $x = 0$, $f(x) = 0$, and never decreases when x is positive, so that for all positive values of x, $f(x)$ is positive.

At a point of inflexion, $f'(x)$ has its least (or greatest) value. Here, the least value of $f'(x)$ is 1, when $x = 0$, so $(0, 0)$ is the point of inflexion. A sketch of the graph shows this point of inflexion:

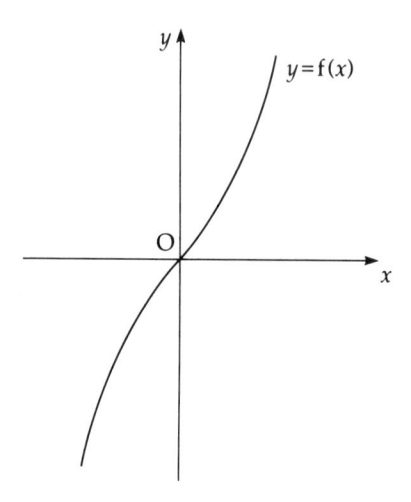

Notes

1. 'Positive' means 'greater than 0'. $f(0) = 0$, but $f'(0) = 4$, so $f(x)$ is increasing and must remain positive.
2. 'Points of inflexion'. At a point of inflexion, $f'(x)$ has a maximum or a minimum value, so $f''(x) = 0$. The converse is not necessarily true, for if $f(x) = x^4$, $f''(x) = 12x^2$, and $f''(0) = 0$, yet $x = 0$ is a minimum.

DIFFERENTIATION

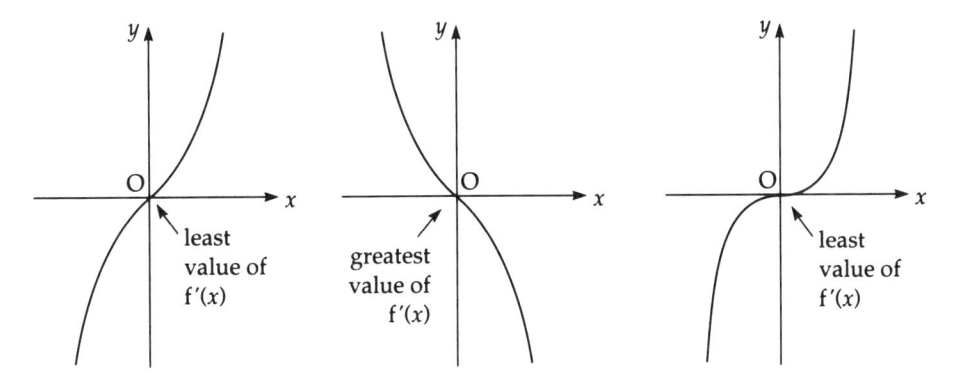

Point of inflexion when $x=0$, $f''(0)=0$

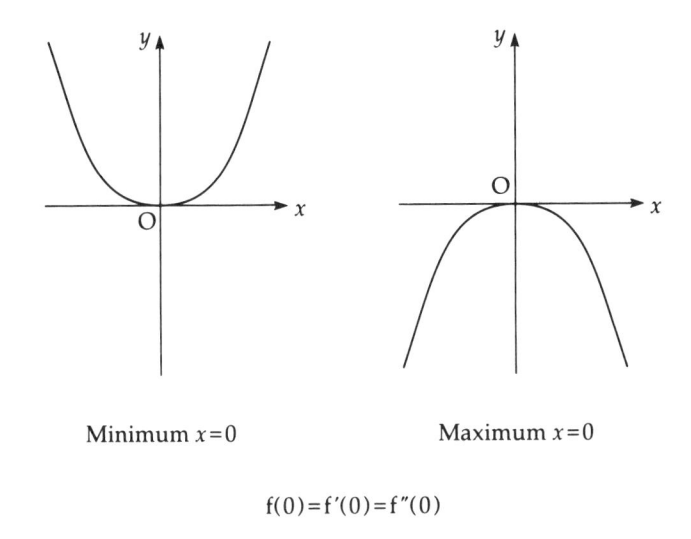

Minimum $x=0$

Maximum $x=0$

$f(0)=f'(0)=f''(0)$

14.5 Derivative of an exponential; second derivative

Question

Given that
$$y = e^{-x^2},$$
find expressions for

(i) $\dfrac{dy}{dx}$,

(ii) $\dfrac{d^2y}{dx^2}$.

equal to zero. Show that these are both points of inflexion.

(WJEC)

Solution

Since $y = e^{-x^2}$

$$\frac{dy}{dx} = -2xe^{-x^2}$$

and $$\frac{d^2y}{dx^2} = (-2)e^{-x^2} + (-2x)(-2x)e^{-x^2}$$

$$= e^{-x^2}(-2 + 4x^2)$$

$$= 2(2x^2 - 1)e^{-x^2}.$$

When $\frac{d^2y}{dx^2} = 0$, $2x^2 - 1 = 0$, since e^{-x^2} is never zero

$$\therefore \quad x^2 = \tfrac{1}{2},$$

$$x = \pm 1/\sqrt{2}.$$

To show these are points of inflexion, it is sufficient to consider just one of them, for we can deduce the other by symmetry. At a point of inflexion $\frac{d^2y}{dx^2}$ must be zero *and must change sign*. Considering values of x near to $1/\sqrt{2}$, approximately 0.707,

when $x = 0.7$, d^2y/dx^2 is negative,
when $x = 1/\sqrt{2}$, $d^2y/dx^2 = 0$,
when $x = 0.8$, d^2y/dx^2 is positive,

so that d^2y/dx^2 changes sign, and $x = 1/\sqrt{2}$ is a point of inflexion. Considering the equation $y = e^{-x^2}$, write this as $f(x) = e^{-x^2}$, then

$$f(-x) = f(x)$$

so the curve is symmetrical about the y-axis, and as $x = 1/\sqrt{2}$ gives a point of inflexion, $x = -1/\sqrt{2}$ will also give a point of inflexion, ie both are points of inflexion.

14.6 Equation of a tangent to a curve; derivative of a quotient

Question

A curve is defined for $x \neq 1$ by the equation

DIFFERENTIATION

$$y = \frac{x+2}{(x-1)^2}.$$

Find the equation of the tangent to the curve at the point where $x = 0$.
Find the coordinates of the point where this tangent crosses the curve again.

(AEB)

Solution

Since $y = \dfrac{x+2}{(x-1)^2}$,

differentiating as a quotient,

$$\begin{aligned}\frac{dy}{dx} &= \frac{(x-1)^2(1) - (x+2)(2)(x-1)}{(x-1)^4} \\ &= \frac{(x-1)[x-1-2(x+2)]}{(x-1)^4} \\ &= \frac{-x-5}{(x-1)^3}.\end{aligned}$$

When $x = 0$, $y = 2$ and $dy/dx = 5$
so the equation of the tangent at $(0, 2)$ is

$$y - 2 = 5(x - 0),$$

ie
$$\underline{\underline{y = 5x + 2}}.$$

To find where the tangent crosses the curve again, we solve the equation of the tangent and the equation of the curve as simultaneous equations,

so $$5x + 2 = \frac{x+2}{(x-1)^2}$$

ie $$(x-1)^2(5x+2) = x + 2.$$

Although this is a cubic, we know that there are two roots $x = 0$, since the tangent touches the curve at two coincident points, so it should be easy to solve the equation once we have simplified it,

$$(x-1)^2(5x+2) = x+2$$
$$\Rightarrow (x^2 - 2x + 1)(5x+2) = x+2$$
$$\Rightarrow 5x^3 - 8x^2 + x + 2 = x + 2$$
$$\Rightarrow 5x^3 - 8x^2 = 0,$$
$$\Rightarrow x^2(5x - 8) = 0,$$
$$x = 0, \text{ (twice) or } x = 8/5.$$

105

Thus the point at which the tangent meets the curve again is when $x = 8/5$, ie (8/5, 10).

Comment

- Looking at the original equation, we can see that $x + 2 = x - 1 + 3$ so the equation can be written

$$y = \frac{x-1}{(x-1)^2} + \frac{3}{(x-1)^2}$$
$$= (x-1)^{-1} + 3(x-1)^{-2}$$

and $\quad dy/dx = (-1)(x-1)^{-2} + 3(-2)(x-1)^{-3}$

possibly a little easier than differentiating as a quotient. Our graphics calculator shows the graph of the curve, enabling us to interpret our results.

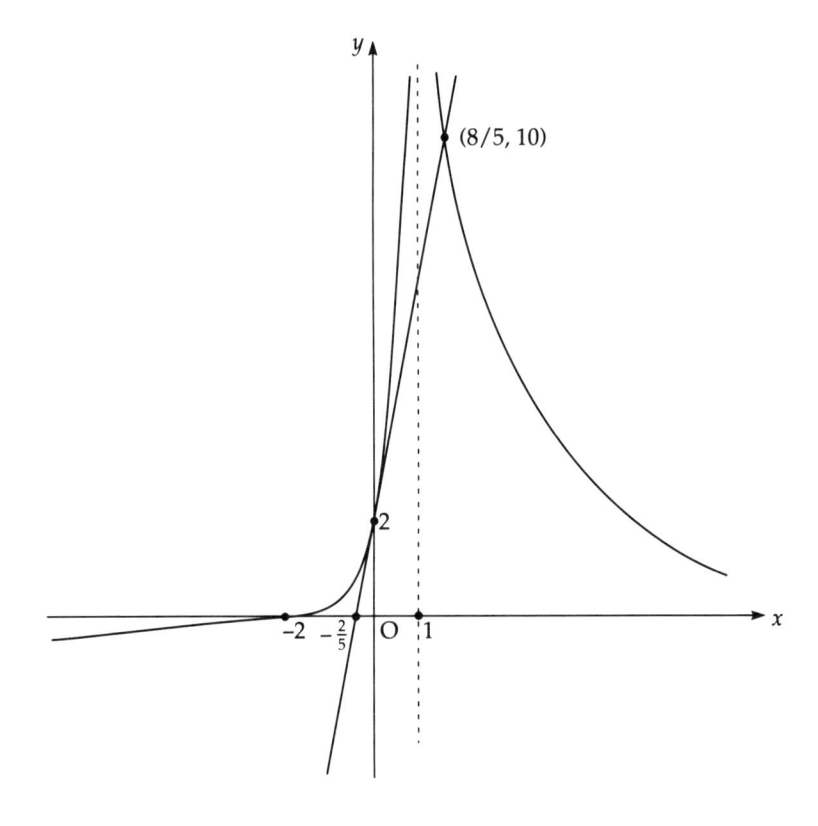

15 INTEGRATION

15.1 Notes
15.2 Deducing an integral from a derived function
15.3 Integration by inspection; by parts; by substitution
15.4 Use of partial fractions; inverse trig and logarithmic forms
15.5 Gradients and areas
15.6 Minimum point; area of a region
15.7 Area of a region; volume of solid of revolution

15.1 Notes

Integrals

$f(x)$ or y	$\int f(x)\,dx$ or $\int y\,dx$
x^n, $n \neq -1$	$\dfrac{1}{n}x^{n+1} + C$
$(ax+b)^n$, $n \neq -1$	$\dfrac{1}{a(n+1)}(ax+b)^{n+1} + C$
$\sin(ax+b)$	$-\dfrac{1}{a}\cos(ax+b) + C$
$\cos(ax+b)$	$\dfrac{1}{a}\sin(ax+b) + C$
$\dfrac{1}{ax+b}$	$\dfrac{1}{a}\ln(ax+b) + C$
$\dfrac{F'(x)}{F(x)}$	$\ln F(x) + C$
e^{ax}	$\dfrac{1}{a}e^{ax} + C$
$\dfrac{1}{\sqrt{a^2 - x^2}}$	$\operatorname{invsin}\left(\dfrac{x}{a}\right) + C$
$\dfrac{1}{a^2 + x^2}$	$\dfrac{1}{a}\operatorname{invtan}\left(\dfrac{x}{a}\right) + C$

107

MATHEMATICS REVISION WORKBOOK

Trigonometric identities are often useful, eg

$$\int \sin^2 \theta \, d\theta = \int \tfrac{1}{2}(1 - \cos 2\theta) \, d\theta$$
$$= \tfrac{1}{2}\theta - \tfrac{1}{4} \sin 2\theta + C,$$

$$\int \sin 3\theta \cos \theta \, d\theta = \int \tfrac{1}{2}(\sin 4\theta + \sin 2\theta) \, d\theta$$
$$= -\tfrac{1}{8} \cos 4\theta - \tfrac{1}{4} \cos 2\theta + C.$$

Areas and volumes

If a region R_1 is bounded by the x-axis, the curve $y = f(x)$ and the lines $x = a$ and $x = b$, the area of R_1 is $\int_a^b y \, dx$.

If the region R_1 is rotated completely about the x-axis, the volume of the solid so formed is $\int_a^b \pi y^2 \, dx$.

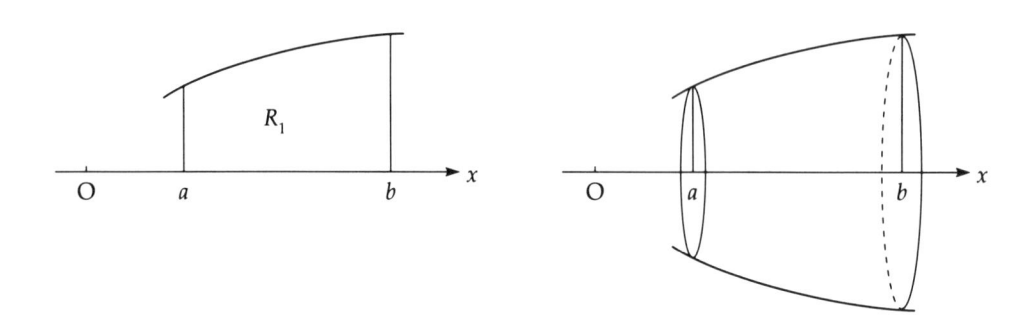

If a region R_2 is bounded by the y-axis, the curve $y = f(x)$ and the lines $y = c$ and $y = d$, the area of R_2 is $\int_c^d x \, dy$.

If the region R_2 is rotated completely about the y-axis, the volume of the solid so formed is $\int_c^d \pi x^2 \, dy$.

INTEGRATION

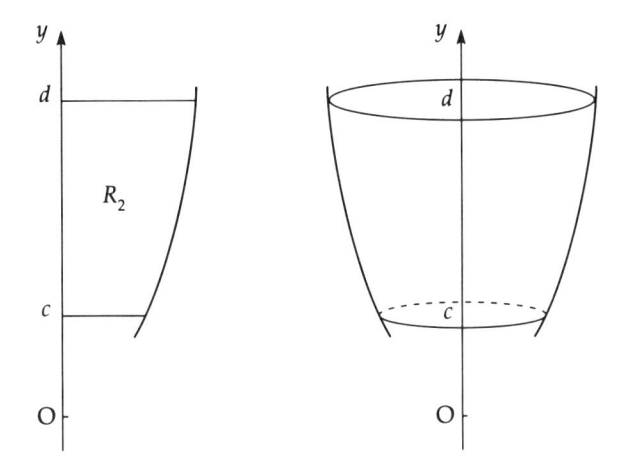

15.2 Deducing an integral from a derived function

Question

(a) Differentiate $(1 + x^3)^{\frac{1}{2}}$ with respect to x.
(b) Use the result from (a), or an appropriate substitution, to find the value of

$$\int_0^2 \frac{x^2}{\sqrt{(1+x^3)}}\,dx.$$

(WJEC)

Solution

(a) This is a 'function of a function', so we differentiate first 'the outside part', then the 'inside part', ie

$$\frac{d}{dx}(1+x^3)^{\frac{1}{2}} = \frac{1}{2}(1+x^3)^{-\frac{1}{2}} \times (3x^2)$$

$$= \underline{\frac{3x^2}{2(1+x^3)^{\frac{1}{2}}}}.$$

(b) Comparing this expression with the derivative that we found in (a), we see that since

109

MATHEMATICS REVISION WORKBOOK

$$\frac{d}{dx}(1+x^3)^{\frac{1}{2}} = \frac{3x^2}{2(1+x^3)^{\frac{1}{2}}},$$

$$\int \frac{x^2}{(1+x^3)^{\frac{1}{2}}}\,dx = \frac{2}{3}(1+x^3)^{\frac{1}{2}}$$

and

$$\int_0^2 \frac{x^2}{(1+x^3)^{\frac{1}{2}}}\,dx = \left[\frac{2}{3}(1+x^3)^{\frac{1}{2}}\right]_0^2$$

$$= \frac{2}{3}(9)^{\frac{1}{2}} - \frac{2}{3}(1)$$

$$= 1\tfrac{1}{3}.$$

Comment

To use an appropriate substitution, write

$$z = 1 + x^3,$$

$$\therefore \quad \frac{dz}{dx} = 3x^2$$

and

$$\int_{x=0}^{x=2} \frac{x^2}{\sqrt{(1+x^3)}}\,dx = \int \frac{1}{\sqrt{(1+x^3)}}(x^2)\,dx$$

$$= \int_{x=0}^{x=2} \frac{1}{z^{\frac{1}{2}}}\left(\frac{1}{3}\frac{dz}{dx}\right)dx$$

$$= \int_{x=0}^{x=2} \frac{1}{3}z^{-\frac{1}{2}}\,dz$$

$$= \left[\frac{2}{3}z^{\frac{1}{2}}\right]_{x=0}^{x=2}$$

$$= \left[\frac{2}{3}(1+x^3)^{\frac{1}{2}}\right]_0^2.$$

We can now evaluate at the upper and lower limits as before, or we could have changed the limits, writing

when $x = 2$, $z = 1 + 2^3 = 9$,

when $x = 0$, $z = 1 + 0^3 = 1$,

so

$$\int_{x=0}^{x=2} \frac{x^2}{\sqrt{(1+x^3)}}\,dx = \left[\frac{2}{3}z^{\frac{1}{2}}\right]_{z=1}^{z=9}$$

$$= \frac{2}{3}(3) - \frac{2}{3}(1)$$

$$= 1\tfrac{1}{3}, \text{ as before.}$$

INTEGRATION

15.3 Integration by inspection; by parts; by substitution

Question

(a) Find the exact value of

(i) $\int_0^3 \frac{x}{1+x^2}\,dx$

(ii) $\int_0^1 x^2\,e^{4x}\,dx.$

(b) Using the substitution $y = \dfrac{1}{x}$, or otherwise, find the exact value of

$$\int_{2/\sqrt{3}}^{2} \frac{1}{x\sqrt{(x^2-1)}}\,dx.$$

Hint We could use the substitution $z = 1 + x^2$, but it is worth looking first to see if the numerator is the derivative of the denominator, since

$$\int \frac{f'(x)}{f(x)}\,dx = \ln f(x).$$

Solution

(a) (i) $\int_0^3 \dfrac{x}{1+x^2}\,dx = \dfrac{1}{2}\int_0^3 \dfrac{2x}{1+x^2}\,dx$

$= \left[\tfrac{1}{2}\ln(1+x^2)\right]_0^3,$ since $\dfrac{d}{dx}(1+x^2) = 2x$

$= \tfrac{1}{2}\ln 10,$ as $\ln 1 = 0.$

(ii) Integrating by parts, we use the formula

$$\int u\frac{dv}{dx} = uv - \int v\frac{du}{dx}$$

where $u = x^2,\ \dfrac{dv}{dx} = e^{4x},$ so $\dfrac{du}{dx} = 2x,\ v = \tfrac{1}{4}e^{4x}$

so $\int x^2 e^{4x}\,dx = x^2\left(\tfrac{1}{4}e^{4x}\right) - \int (2x)\left(\tfrac{1}{4}e^{4x}\right)dx$

$= \tfrac{1}{4}x^2 e^{4x} - \tfrac{1}{2}\int x e^{4x}\,dx.$ (1)

111

MATHEMATICS REVISION WORKBOOK

Now we use integration by parts again, taking

$$u = x, \quad \frac{dv}{dx} = e^{4x}, \text{ so } \frac{du}{dx} = 1, \quad v = \tfrac{1}{4}e^{4x}$$

and $\displaystyle\int xe^{4x}\, dx = x(\tfrac{1}{4}e^{4x}) - \int (1)(\tfrac{1}{4}e^{4x})dx$

$$= \tfrac{1}{4}xe^{4x} - \tfrac{1}{4}\int e^{4x}\, dx$$

$$= \tfrac{1}{4}xe^{4x} - \tfrac{1}{16}e^{4x}.$$

Substituting in (1), $\displaystyle\int x^2 e^{4x}\, dx = \tfrac{1}{4}x^2 e^{4x} - \tfrac{1}{2}(\tfrac{1}{4}xe^{4x} - \tfrac{1}{16}e^{4x})$

$$= (\tfrac{1}{4}x^2 - \tfrac{1}{8}x + \tfrac{1}{32})e^{4x}$$

and $\displaystyle\int_0^1 x^2 e^{4x}\, dx = \left[(\tfrac{1}{4}x^2 - \tfrac{1}{8}x + \tfrac{1}{32})e^{4x}\right]_0^1$

$$= \underline{\underline{\tfrac{1}{32}(5e^4 - 1)}}.$$

Comment

- When using integration by parts, choose the function to be u so that if possible $\dfrac{du}{dx}$ is simpler than u, or at worst no more difficult. Here $2x$ is one degree less than x^2, and $\tfrac{1}{4}e^{4x}$ is no more difficult to integrate than e^{4x}. We realise that we shall have to carry out integration by parts twice, as $\dfrac{d}{dx}(x^2) = 2x$, $\dfrac{d}{dx}(2x) = 2$.

(b) *Hint* The expression to be integrated looks awkward, but we realise that $\dfrac{d}{dx}\left(\dfrac{1}{x}\right) = -\dfrac{1}{x^2}$, and $\sqrt{(x^2 - 1)} = x\sqrt{(1 - \dfrac{1}{x^2})}$, so we can substitute fairly easily, replacing $\dfrac{1}{x}$ by y.

Solution

To save writing down the limits each time, treat the integral first as an indefinite integral, so

$$\int \frac{1}{x\sqrt{(x^2 - 1)}}\, dx = \int \frac{1}{x^2\sqrt{\left(\dfrac{1}{x^2} - 1\right)}}\, dx$$

$$= \int \frac{1}{\sqrt{\left(\dfrac{1}{x^2} - 1\right)}} \cdot \frac{1}{x^2}\, dx.$$

Now if $y = \dfrac{1}{x}$, $\dfrac{dy}{dx} = -\dfrac{1}{x^2}$

112

INTEGRATION

so $$\int \frac{1}{\sqrt{\left(\frac{1}{x^2} - 1\right)}} \cdot \frac{1}{x^2} \cdot dx = \int \frac{1}{\sqrt{(1-y^2)}} (-dy)$$

$$= -\sin^{-1} y.$$

Changing the limits of integration,

when $x = 2$, $y = \frac{1}{2}$,

when $x = \frac{2}{\sqrt{3}}$, $y = \frac{1}{2}\sqrt{3}$,

so $$\int_{x=\frac{2}{\sqrt{3}}}^{x=2} \frac{1}{x\sqrt{(x^2-1)}} dx = \left[-\sin^{-1} y\right]_{y=\frac{1}{2}\sqrt{3}}^{y=\frac{1}{2}}$$

$$= -\left(\frac{\pi}{6} - \frac{\pi}{3}\right)$$

$$= \underline{\underline{\frac{\pi}{6}}}.$$

Comment

There are several other substitutions we could have used. If we try

$$\sec \theta = x, \quad \sec \theta \tan \theta \frac{d\theta}{dx} = 1$$

we reach $$\int 1 \, d\theta = \theta$$

$$= \sec^{-1} x$$

whence the value of the integral is $\left[\sec^{-1} x\right]_{2/\sqrt{3}}^{2}$

ie $\frac{\pi}{3} - \frac{\pi}{6}$, ie $\frac{\pi}{6}$.

15.4 Use of partial fractions; inverse trig and logarithmic forms

Questions

Given $f(x) = \frac{1}{(x+2)(x^2+4)}$, express $f(x)$ in partial fractions.

Show that

$$\int_0^2 f(x) \, dx = \frac{(\pi + 2 \ln 2)}{32}.$$

(L)

Solution

Write
$$\frac{1}{(x+2)(x^2+4)} = \frac{A}{x+2} + \frac{Bx+C}{x^2+4},$$

$$\frac{1}{(x+2)(x^2+4)} = \frac{A(x^2+4) + (Bx+C)(x+2)}{(x+2)(x^2+4)}.$$

Equating the numerators of the fractions,

$$1 \equiv A(x^2+4) + (Bx+C)(x+2), \qquad (1)$$

ie
$$1 \equiv (A+B)x^2 + (2B+C)x + 4A + 2C. \qquad (2)$$

Putting $x = -2$ in (1),

$1 = 8A$, ie $A = \frac{1}{8}$.

Equating the coefficients of x^2 in (2),

$0 = A + B$, ie $B = -\frac{1}{8}$, as $A = \frac{1}{8}$.

Equating the coefficients of x in (2),

$0 = 2B + C$, ie $C = \frac{1}{4}$, as $B = -\frac{1}{8}$.

so
$$\frac{1}{(x+2)(x^2+4)} = \frac{\frac{1}{8}}{x+2} + \frac{-\frac{1}{8}x + \frac{1}{4}}{x^2+4}$$

$$= \frac{1}{8(x+2)} + \frac{2-x}{8(x^2+4)}.$$

To find $\int f(x)dx$, recall that

$$\int \frac{1}{x+a}dx = \ln(x+a), \quad \int \frac{x}{x^2+a^2}dx = \tfrac{1}{2}\ln(x^2+a^2), \quad \int \frac{1}{x^2+a^2}dx = \frac{1}{a}\tan^{-1}\frac{x}{a}$$

so
$$\int \frac{1}{(x+2)}dx = \ln(x+2), \quad \int \frac{x}{x^2+4}dx = \tfrac{1}{2}\ln(x^2+4)$$

and
$$\int \frac{1}{x^2+4}dx = \tfrac{1}{2}\tan^{-1}\frac{x}{2}.$$

INTEGRATION

Now
$$\int_0^2 f(x) \, dx = \int_0^2 \left[\frac{1}{8(x+2)} + \frac{2}{8(x^2+4)} - \frac{x}{8(x^2+4)} \right] dx$$
$$= \left[\frac{1}{8} \ln(x+2) + \left(\frac{2}{8}\right)\left(\frac{1}{2} \tan^{-1}\left(\frac{x}{2}\right)\right) - \left(\frac{1}{8}\right)\left(\frac{1}{2}\right) \ln(x^2+4) \right]_0^2$$
$$= \left[\frac{1}{8} \ln 4 - \frac{1}{16} \ln 8 + \frac{1}{8}\left(\frac{\pi}{4}\right) \right] - \left[\frac{1}{8} \ln 2 - \frac{1}{16} \ln 4 \right]$$
$$= \frac{1}{16} \ln 2 + \frac{\pi}{32}$$
$$= \frac{(\pi + 2 \ln 2)}{32}.$$

Comment

- Writing $\ln 8 = 3 \ln 2$, $\ln 4 = 2 \ln 2$, and using $\tan^{-1}(1) = \frac{1}{4}\pi$.

15.5 Gradients and areas

Question

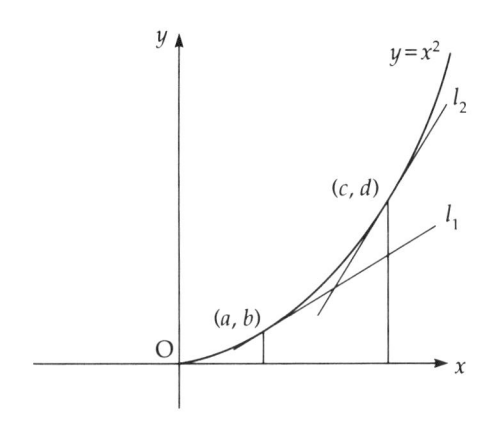

The figure shows the curve with equation $y = x^2$, $x > 0$, and two lines l_1 and l_2 which are tangents to the curve at (a, b) and (c, d) respectively. The line l_1 has gradient 1 and the line l_2 has gradient 2.

(a) Find the values of a, b, c and d.
(b) Find the area of the finite region bounded by the curve with equation $y = x^2$, the line $x = a$, the line $x = c$ and the x-axis.

(L)

Solution

(a) The gradient of the curve $y = x^2$ is given by $dy/dx = 2x$, so that since the gradient at (a, b) is 1, $2a = 1$, $a = \frac{1}{2}$.

115

Since (a, b) lies on $y = x^2$ and $a = \frac{1}{2}$, $b = \frac{1}{4}$.
Since the gradient at (c, d) is 2, $2c = 2$, $c = 1$.
Since (c, d) also lies on $y = x^2$ and $c = 1$, $d = 1$,

so $\qquad a = \frac{1}{2}, \; b = \frac{1}{4}, \; c = 1 \text{ and } d = 1.$

(b) The region required is that shaded in the figure. Its area A is given by

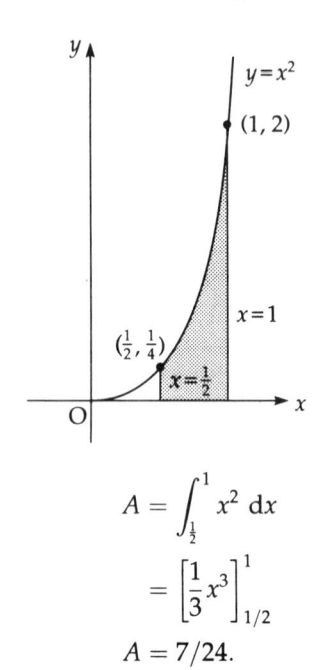

$$A = \int_{\frac{1}{2}}^{1} x^2 \, dx$$

$$= \left[\frac{1}{3} x^3 \right]_{1/2}^{1}$$

$$A = 7/24.$$

15.6 Minimum point; area of a region

Question

The figure shows a sketch of the curve defined for $x > 0$ by the equation

$$y = x^2 \ln x.$$

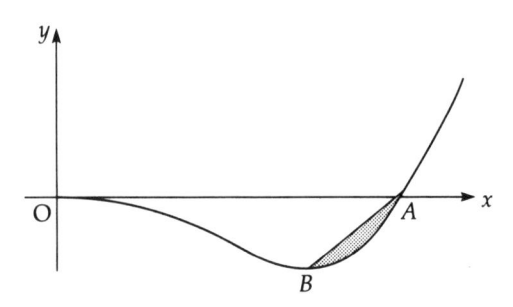

INTEGRATION

The curve crosses the x-axis at A and has a local minimum at B.

(a) State the coordinates of A and calculate the gradient of the curve at A.

(b) Determine the coordinates of B and determine the value of $\dfrac{d^2y}{dx^2}$ at B.

(c) The region bounded by the line segment AB and an arc of the curve is R, as shaded in the figure. Show that the area of R is

$$\frac{1}{36}(4 - e^{-\frac{3}{2}} - 9e^{-1}).$$

(AEB)

Hints Remember $\ln x = 0 \Leftrightarrow x = 1$,

$\ln x = a \Leftrightarrow x = e^a$

$\ln(e^b) = b$; in particular, $\ln(e^{-\frac{1}{2}}) = -\frac{1}{2}$.

Also $\int x^n \ln x \, dx$ is integrated by parts, integrating x^n and differentiating $\ln x$.

Solution

(a) When $y = 0$, $x^2 \ln x = 0$, $\therefore x^2 = 0$ or $\ln x = 0$.

When $x^2 = 0$, $x = 0$ (twice), showing the curve touches the x-axis at the origin; when $\ln x = 0$, $x = 1$, so the coordinates of A are $(1, 0)$.

Since $y = x^2 \ln x$, $dy/dx = 2x \ln x + x^2 \left(\dfrac{1}{x}\right)$.

When $x = 1$, $dy/dx = 1$, since $\ln 1 = 0$, so

the gradient of the curve at A is 1.

(b) As B is a local minimum, the gradient at B is zero,

ie $\qquad\qquad 2x \ln x + x = 0,$

ie $\qquad\qquad x(2 \ln x + 1) = 0.$

The solution $x = 0$ clearly refers to the origin, so that B,

$$2 \ln x + 1 = 0,$$
$$\ln x = -\tfrac{1}{2},$$
$$x = e^{-\frac{1}{2}}.$$

Since the equation of the curve is $y = x^2 \ln x$, the y coordinate of B is

$$[e^{-\frac{1}{2}}]^2 \, [-\tfrac{1}{2}],$$

ie $\qquad\qquad -\tfrac{1}{2}e^{-1},$

and B is $\qquad (e^{-\frac{1}{2}}, -\tfrac{1}{2}e^{-1}).$

117

MATHEMATICS REVISION WORKBOOK

Since
$$\frac{dy}{dx} = 2x \ln x + x,$$
$$\frac{d^2y}{dx^2} = 2 \ln x + 2x\left(\frac{1}{x}\right) + 1$$
$$= 2 \ln x + 3$$

so that since $\ln x = -\frac{1}{2}$ at B,
$$\frac{d^2y}{dx^2} = 2.$$

(c) The area of R' is given by
$$\int_{e^{-1/2}}^{1} x^2 \ln x \, dx.$$

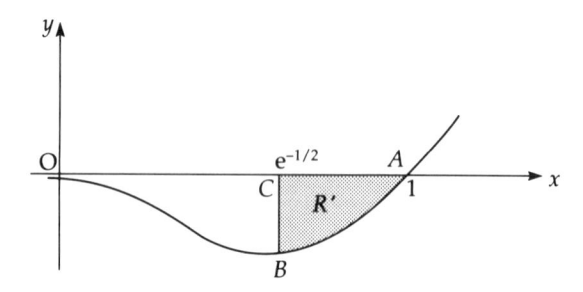

Integrate $x^2 \ln x$ by parts, taking
$$u = \ln x \Rightarrow \frac{du}{dx} = \frac{1}{x}$$
and $\quad \frac{dv}{dx} = x^2 \Rightarrow v = \frac{1}{3}x^3,$

so $\displaystyle\int x^2 \ln x \, dx = \frac{1}{3}x^3 \ln x - \int \frac{1}{3}x^3 \left(\frac{1}{x}\right) dx$
$$= \frac{1}{3}x^3 \ln x - \int \frac{1}{3}x^2 \, dx$$
$$= \frac{1}{3}x^3 \ln x - \frac{1}{9}x^3.$$

INTEGRATION

Thus $\int_{e^{-1/2}}^{1} x^2 \ln x \, dx = \left[\frac{1}{3} x^3 \ln x - \frac{1}{9} x^3 \right]_{e^{-\frac{1}{2}}}^{1}$

$= \left[0 - \frac{1}{9} \right] - \left[\frac{1}{3} e^{-\frac{3}{2}} \left(-\frac{1}{2} \right) - \frac{1}{9} e^{-\frac{3}{2}} \right]$

$= -\frac{1}{9} + \frac{5}{18} e^{-\frac{3}{2}}.$

This is negative, as expected from the graph, so the area of the region bounded by the curve, the x-axis and the line BC is

$$\frac{1}{9} - \frac{5}{18} e^{-\frac{3}{2}}$$

The area of the triangle ABC, from the diagram is

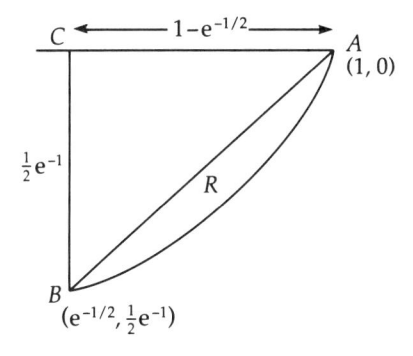

$$\frac{1}{2} \left(\frac{1}{2} e^{-1} \right) \left(1 - e^{-\frac{1}{2}} \right) = \frac{1}{4} e^{-1} - \frac{1}{4} e^{-\frac{3}{2}}$$

and the area of R is

$$\frac{1}{9} - \frac{5}{18} e^{-\frac{3}{2}} - \left[\frac{1}{4} e^{-1} - \frac{1}{4} e^{-\frac{3}{2}} \right]$$

$$= \underline{\underline{\frac{1}{36} \left(4 - e^{-\frac{3}{2}} - 9 e^{-1} \right)}}$$

15.7 Area of a region; volume of solid of revolution

Question

The region R is bounded by the x-axis and the part of the curve $y = \sin 2x$ between $x = 0$ and $x = \frac{1}{2}\pi$. Use integration to find the exact values of

(i) the area of R,
(ii) the volume of the solid formed when R is rotated completely about the x-axis.

(C)

Solution

(i)

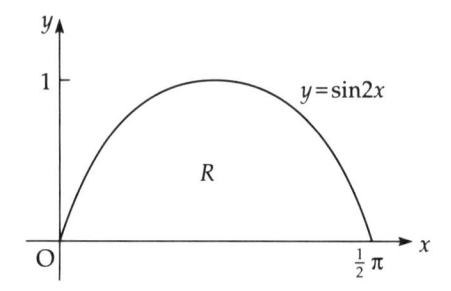

Our graphics calculator enables us to see the shape of the region R. The area A of R is

$$\int_0^{\frac{1}{2}\pi} y \, dx, \text{ ie } \int_0^{\frac{1}{2}\pi} \sin 2x \, dx$$

$$= \left[-\frac{1}{2} \cos 2x \right]_0^{\frac{1}{2}\pi}$$

$$= -\frac{1}{2}(-1) - \left(-\frac{1}{2} \times 1\right)$$

$$= 1.$$

Check

From the sketch, A is certainly less than the area of a rectangle, length $\frac{1}{2}\pi$, height 1, which is about 1.5. It is greater than the area of a triangle on the base $\frac{1}{2}\pi$, height 1, ie greater than $\frac{1}{4}\pi$, about 0.75, so that a value of 1 is reasonable. We may remember that the region bounded by the sine curve and the x-axis from 0 to π is 2.

Solution

(ii) The volume V of the solid of revolution is $\int y^2 \, dx$, ie

$$V = \pi \int_0^{\frac{1}{2}\pi} \sin^2 2x \, dx.$$

Write $\sin^2 2x = \frac{1}{2}(1 - \cos 4x)$, using $\cos 2\theta = 1 - 2\sin^2 \theta$

so
$$V = \pi \int_0^{\frac{1}{2}\pi} \frac{1}{2}(1 - \cos 4x) dx$$

$$= \pi \left[\frac{1}{2}x - \frac{1}{8}\sin 4x \right]_0^{\frac{1}{2}\pi}$$

$$= \pi \left(\frac{1}{4}\pi \right)$$

$$= \frac{1}{4}\pi^2.$$

Check

The solid is contained inside a cylinder, length $\frac{1}{2}\pi$, base radius 1, whose volume is $\frac{1}{2}\pi^2$. This value of $\frac{1}{4}\pi^2$ is clearly reasonable.

16 DIFFERENTIAL EQUATIONS

16.1 Notes
16.2 Integration by parts; differential equation with separable variables
16.3 Separable variables; inverse trig form
16.4 Tangent field; family of curves

16.1 Notes

Forming differential equation

'Rate of change of y with respect to x' is written $\dfrac{dy}{dx}$,

eg the rate of change of a volume V with respect to time t is

    (i) proportional to t,

$$\text{means } \frac{dV}{dt} = kt,$$

    (ii) proportional to t^2,

$$\text{means } \frac{dV}{dt} = kt^2,$$

    (iii) inversely proportional to t,

$$\text{means } \frac{dV}{dt} = \frac{k}{t},$$

k being a constant in each case.

Solving differential equations

If $\dfrac{dy}{dx} = f(x)$, $y = \displaystyle\int f(x)\, dx$,

eg if $\dfrac{dy}{dx} = \cos x$, $y = \sin x + C$.

If $\dfrac{dy}{dx} = f(y)\, F(x)$, $\displaystyle\int \frac{1}{f(y)}\, dy = \int F(x)\, dx$

eg if $\dfrac{dy}{dx} = y^2 x^3$, $\dfrac{1}{y^2}\dfrac{dy}{dx} = x^3$,

$$\int \frac{1}{y^2}\, dy = \int x^3\, dx,$$

$$-\frac{1}{y} = \tfrac{1}{4}x^4 + C.$$

122

DIFFERENTIAL EQUATIONS

Tangent-fields

$\frac{dy}{dx}$ describes the gradient of each curve in the family,

eg $\frac{dy}{dx} = x$ says that the gradient is $-2, -1, 0, 1, 2$ when $x = -2, -1, 0, 1, 2$.

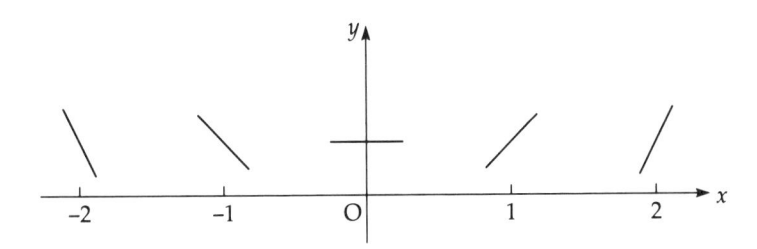

so the family of curves is

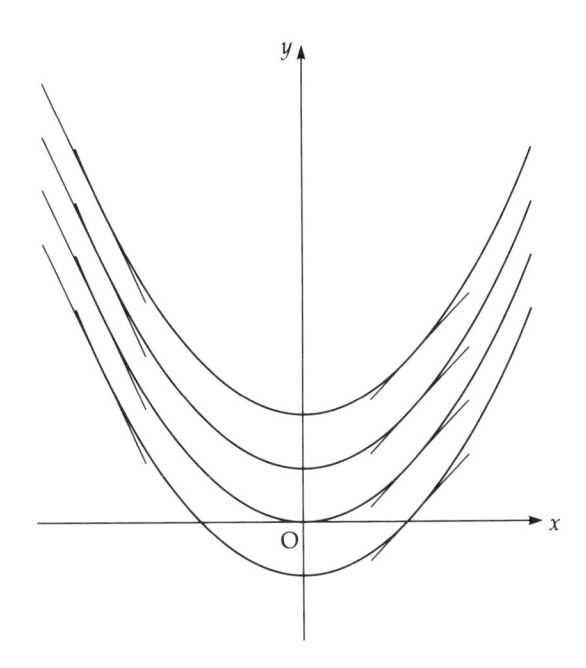

MATHEMATICS REVISION WORKBOOK

16.2 Integration by parts; differential equation with separable variables

Question

(a) Use integration by parts to find
$$\int x^{\frac{1}{2}} \ln x \, dx$$

(b) Find the solution of the differential equation
$$\frac{dy}{dx} = (xy)^{\frac{1}{2}} \ln x$$
for which $y = 1$ when $x = 1$.

(AEB)

(a) Solution

Since $\ln x$ differentiates easily to give $\frac{1}{x}$, we take $x^{\frac{1}{2}}$ as $\frac{dv}{dx}$ and $\ln x$ as u in the formula for integrating by parts, so

$$u = \ln x \Rightarrow \frac{du}{dx} = \frac{1}{x}$$

and
$$\frac{dv}{dx} = x^{\frac{1}{2}} \Rightarrow v = \tfrac{2}{3} x^{\frac{3}{2}}$$

Thus
$$\int u \frac{dv}{dx} \, dx = uv - \int v \frac{du}{dx} \, dx$$

$$\Rightarrow \int (\ln x) \, x^{\frac{1}{2}} \, dx = (\tfrac{2}{3} x^{\frac{3}{2}})(\ln x) - \int \frac{1}{x} (\tfrac{2}{3} x^{\frac{3}{2}}) \, dx$$

$$= \tfrac{2}{3} x^{\frac{3}{2}} \ln x - \int \tfrac{2}{3} x^{\frac{1}{2}} \, dx$$

$$= \underline{\underline{\tfrac{2}{3} x^{\frac{3}{2}} \ln x - \tfrac{4}{9} x^{\frac{3}{2}} + C}}.$$

(b) Solution

Separating the variables, we have

$$\frac{1}{y^{\frac{1}{2}}} \frac{dy}{dx} = x^{\frac{1}{2}} \ln x \quad \text{ie} \quad 2y^{\frac{1}{2}} = \tfrac{2}{3} x^{\frac{3}{2}} \ln x - \tfrac{4}{9} x^{\frac{3}{2}} + C$$

using the result from (a).
But $y = 1$ when $x = 1$, so $2 = -\tfrac{4}{9} + C$,
$C = 2\tfrac{4}{9}$, and the solution is $\underline{y^{\frac{1}{2}} = \tfrac{1}{3} x^{\frac{3}{2}} \ln x - \tfrac{2}{9} x^{\frac{3}{2}} + 1\tfrac{2}{9}}$.
after dividing both sides by 2.

DIFFERENTIAL EQUATIONS

16.3 Separable variables; inverse trig form

Question

Find the solution of the differential equation

$$\frac{dy}{dx} = (1+x^2)(1+y^2)$$

which satisfies $y = 1$ when $x = 0$. Give your answer in the form $y = f(x)$.

(AEB)

Solution

Separating the variables, we have

$$\frac{1}{1+y^2}\frac{dy}{dx} = (1+x^2)$$

ie $$\int \frac{1}{1+y^2} dy = \int (1+x^2) dx$$

ie $\quad \text{invtan } y = x + \tfrac{1}{3}x^3 + C.$

But $y = 1$ when $x = 0$, so $C = \pi/4$, since invtan $1 = \pi/4$, and the solution in the form $y = f(x)$ is

$$y = \tan(x + \tfrac{1}{3}x^3 + \pi/4).$$

16.4 Tangent field; family of curves

Question

A family of curves is defined by the differential equation

$$\frac{dy}{dx} = -\frac{y}{x}.$$

(i) Find the gradients of the curves at (2,2) and (−1,0).

(ii) Copy and complete the diagram below showing the tangent field at points with integer coefficients between −2 and +2, excluding the origin:

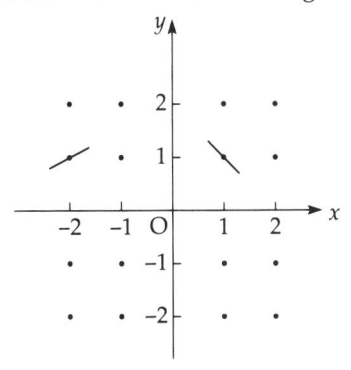

125

(iii) Solve the differential equation assuming that x and y are both positive and give, in simplified form, the equation of the curve from this family which passes through the point (1,1). Sketch this curve on the same diagram as the tangent field.

(iv) Another family of curves is defined by
$$\frac{dy}{dx} = -\frac{(y-2)}{(x+1)}.$$

Draw another sketch showing a typical member of the family for $x > -1$, labelling any important features.

(MEI)

Hint The diagram illustrates the differential equation given at $(-2, 1)$, $dy/dx = -(1)/(-2) = \frac{1}{2}$ as shown, and at (1,1), $dy/dx = -1$, as shown.

Solution

(i) At (2,2), $dy/dx = -2/2 = -1$.
At $(-1,0)$, $dy/dx = -0/1 = 0$.

(ii) First add to the diagram the gradients at (2,2) and $(-1,0)$ which we have just calculated:

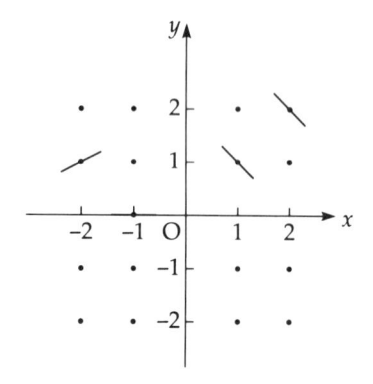

Now choose any one value of x, say $x = 2$, and calculate the values of $-y/x$ at the points along the line $x = 2$

ie at $(2, -2)$, $dy/dx = -(-2)/(2) = 1$;
at $(2, -1)$, $dy/dx = -(-1)/2 = \frac{1}{2}$;
at $(2, 0)$, $dy/dx = -0/2 = 0$;
at $(2, 1)$ $dy/dx = -1/2 = -\frac{1}{2}$;
at $(2, 2)$, we have already found $dy/dx = -1$.

Add these tangent lines to our sketch, and notice that as y increases, the gradient decreases, being positive when $y < 0$, negative when $y > 0$.

DIFFERENTIAL EQUATIONS

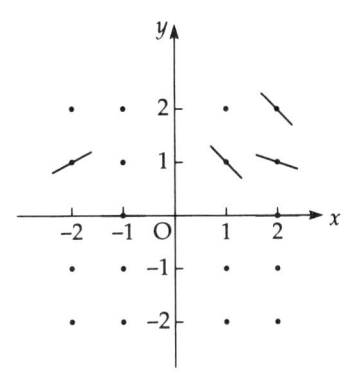

Continue calculating the value of dy/dx at the points with integer coordinates, and mark the gradients as below:

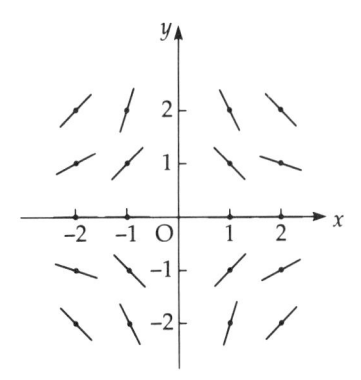

Notice the pattern in the lines, which would show any errors we had made in our calculations. Notice also that the gradient at $(0,0)$ is undefined – we were not asked to find that.

(iii) To solve the differential equation we may notice that when

$$\frac{dy}{dx} = -\frac{y}{x},$$

$$x\frac{dy}{dx} = -y,$$

$$x\frac{dy}{dx} + y = 0,$$

ie $xy = k$, for some constant k.

But this passes through the point $(1,1)$, so $k = 1$, and the equation of the curve of this family through $(1,1)$ is $xy = 1$.

127

We now add to the diagram that part of the curve $xy = 1$ for which x and y are positive:

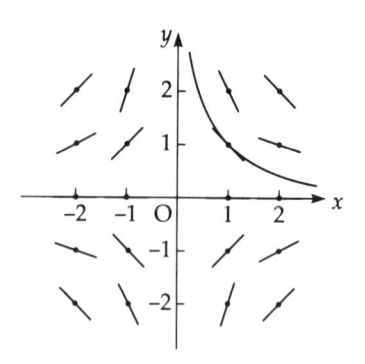

We recognise this as a rectangular hyperbola of the family $xy = c^2$.

(iv) Comparing $\dfrac{dy}{dx} = -\dfrac{y}{x}$ with $\dfrac{dy}{dx} = -\dfrac{(y-2)}{x+1}$

we see that for the first curve, dy/dx is infinite when $x = 0$;
        for the other curve dy/dx is infinite when $x + 1 = 0$, ie $x = -1$.
Similarly, for the first curve $dy/dx = 0$ when $y = 0$,
        for the other curve $dy/dx = 0$ when $y - 2 = 0$, ie $y = 2$.
Whereas the first family of curves have $x = 0$ and $y = 0$ as asymptotes, the second family have $x = -1$ and $y = 2$ as asymptotes, so that a typical member of this family is that given below:

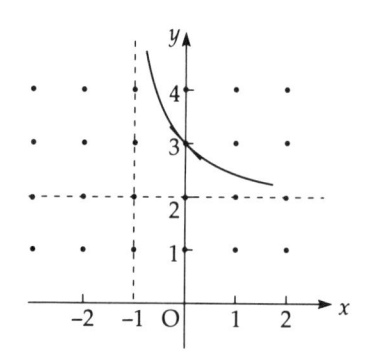

Comment

- If we do not use this method to solve the differential equation, we have to separate the variables,

DIFFERENTIAL EQUATIONS

ie $\quad \dfrac{dy}{dx} = -\dfrac{y}{x},$

$\dfrac{1}{y}\dfrac{dy}{dx} = -\dfrac{1}{x},$

$\ln y = -\ln x + \ln A,$ taking the constant as $\ln A$ instead of $k,$

$\therefore \quad \ln y + \ln x = \ln A,$

$\underline{\underline{xy = A,}}$ as before.

If we do not use $\ln A$, we have

$$\ln y = -\ln x + C$$
$$\ln y + \ln x = C,$$
$$\underline{\underline{xy = e^C.}}$$

17 APPROXIMATIONS TO INTEGRALS

17.1 Notes
17.2 Trapezium rule, given y = f(x)
17.3 Simpson's rule, given the ordinates
17.4 Simpson's rule, given y = f(x)

17.1 Notes

The **trapezium rule** gives an approximation to the area of a region. When this region is bounded by a curve $y = f(x)$, the x-axis and by two ordinates, it can be divided into n strips each of width h, and the area is approximately

$$\tfrac{1}{2}h[y_0 + 2(y_1 + y_2 + y_3 \ldots) + y_n].$$

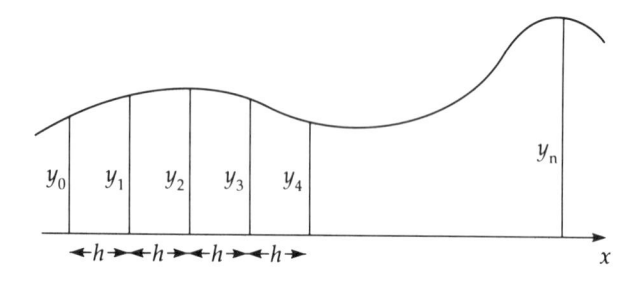

Simpson's rule says, that when n is even, the area is approximately

$$\tfrac{1}{3}h[y_0 + 4y_1 + 2y_2 + 4y_3 + 2y_4 \ldots\ldots + 4y_{n-1} + y_n].$$

17.2 Trapezium rule, given y = f(x)

Question

Use the trapezium rule with intervals of 0.5 to find an approximation for

$$\int_1^{2.5} \frac{1}{1 + \ln x}\, dx$$

giving your answer correct to 2 decimal places.

(C)

APPROXIMATIONS TO INTEGRALS

Solution

As we have intervals of width 0.5, the ordinates are at $x = 1$, 1.5, 2 and 2.5. A rough sketch helps to clarify our ideas:

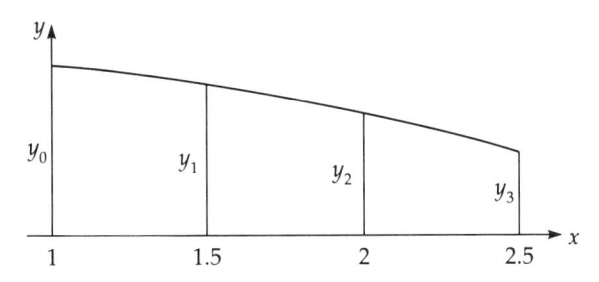

The trapezium rule with three strips (four ordinates) is
$$A \approx \tfrac{1}{2}h[y_0 + 2(y_1 + y_2) + y_3].$$

The value of y_0 is $1/(1 + \ln 1)$, ie 1
of y_1 is $1/(1 + \ln 1.5)$, ie $0.7115\ldots$
of y_2 is $1/(1 + \ln 2)$, ie $0.5906\ldots$
of y_3 is $1/(1 + \ln 2.5)$, ie $0.5218\ldots$

so that $A \approx \tfrac{1}{2} \times 0.5 \times [1 + 2(0.7115 + 0.5906) + 0.5218]$
$\approx 1.0315\ldots$
$= \underline{\underline{1.03}}$, to two decimal places.

Comment

With some practice on our calculator we should find that we do not need to evaluate y_1, y_2 and y_3, but can work out the value of A without any intermediate writing, from

$$A \approx \tfrac{1}{2} \times 0.5 \times \left[1 + 2\left(\frac{1}{1 + \ln 1.5} + \frac{1}{1 + \ln 2}\right) + \frac{1}{1 + \ln 2.5}\right].$$

Use the trapezium rule with intervals of width 0.5 to find an approximation for

$$\int_1^{2.5} \frac{1}{1 + \ln x} \, dx,$$

giving your answer correct to 2 decimal places.

(C)

Alternative solution

The values of x between 1 and 2.5 at intervals of 0.5 are 1, 1.5, 2 and 2.5. Tabulate the corresponding values of the ordinates:

131

MATHEMATICS REVISION WORKBOOK

x	1	1.5	2	2.5
$\dfrac{1}{1+\ln x}$	1	0.7115	0.5906	0.5218

By the trapezium rule,

$$\int_1^{2.5} \frac{1}{1+\ln x}\,dx \approx \tfrac{1}{2}(0.5)[1 + 2(0.7115 + 0.5906) + 0.5218]$$

$$= 1.0315$$

$$= \underline{1.03, \text{ to 2 decimal places.}}$$

Comment

How much working we need to write down will depend on our skill at using our calculator. Denoting $1/(1+\ln x)$ by $f(x)$, we can write

$$\int_1^{2.5} \frac{1}{1+\ln x}\,dx = \tfrac{1}{2}(0.5)[f(1) + 2(f(1.5) + f(2)) + f(2.5)]$$

$$= 1.031522544\ldots$$

$$= 1.03, \text{ to 2 decimal places.}$$

We should not be making our method clear if we wrote down less than this, and anyway it would be easy to make errors.

17.3 Simpson's rule, given the ordinates

Question

The table shows three values of x with the corresponding values of f(x):

x	-1	2	5
f(x)	8	26	206

Use Simpson's rule with 3 ordinates to find an approximate value for

$$\int_{-1}^{5} f(x)\,dx.$$

(L)

Solution

Simpson's rule for 3 ordinates is

$$I \approx \tfrac{1}{3}h(y_0 + 4y_1 + y_2)$$

APPROXIMATIONS TO INTEGRALS

where h is the distance between each pair of ordinates. Here
$$h = 3,\ y_0 = 8,\ y_1 = 26 \text{ and } y_2 = 206, \text{ so}$$
$$I \approx \tfrac{1}{3}(3)(8 + 4 \times 26 + 206)$$
$$= \underline{\underline{318.}}$$

Comment

If we had had 5 ordinates, Simpson's rule of course is
$$I \approx \tfrac{1}{3}h(y_0 + 4y_1 + 2y_2 + 4y_3 + y_4).$$

Similarly for any odd number of ordinates.

17.4 Simpson's rule, given $y = f(x)$

Question

Use Simpson's rule with 5 ordinates and an interval of 0.25 to find an approximation to the integral
$$\int_0^1 \sqrt{x^3 + 1}\ dx.$$

Give your answer correct to 2 decimal places.

(WJEC)

Solution

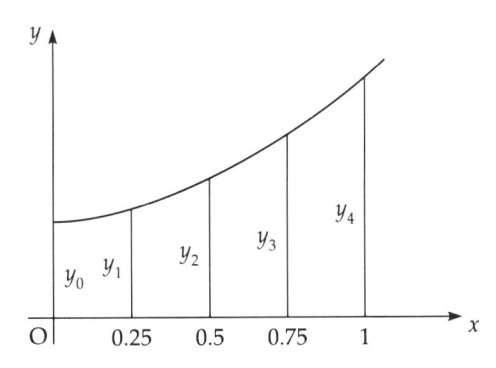

Simpson's rule with 5 ordinates (4 intervals) is
$$I \approx \tfrac{1}{3}h(y_0 + 4y_1 + 2y_2 + 4y_3 + y_4).$$

133

MATHEMATICS REVISION WORKBOOK

Here, $y_0 = \sqrt{(0^3+1)}$, ie 1
$y_1 = \sqrt{(0.25^3+1)}$, ie 1.0078, to 4 dp
$y_2 = \sqrt{(0.5^3+1)}$, ie 1.0607, "
$y_3 = \sqrt{(0.75^3+1)}$, ie 1.1924, "
$y_4 = \sqrt{(1^3+1)}$, ie 1.4142, "

so $I \approx \frac{1}{3}(0.25)[1 + 4 \times 1.0078 + 2 \times 1.0607 + 4 \times 1.1924 + 1.4142]$
   $= 1.11136\ldots$
   $= \underline{\underline{1.11, \text{ to 2 decimal places.}}}$

Comment

We can evaluate this approximation using our calculator without writing down any working, using
$I \approx \frac{1}{3}(0.25)[1 + 4 \times \sqrt{(0.25^3+1)} + 2 \times \sqrt{(0.5^3+1)} + 4 \times \sqrt{(0.75^3+1)} + \sqrt{(1^3+1)}]$
  $= 1.111363\ldots$
  $= 1.11$, to 2 decimal places.

18 ITERATIVE METHODS OF SOLVING EQUATIONS

18.1 Notes

18.2 Locating a root; solving using Newton–Raphson method

18.3 An iteration of the form $x_{r+1} = f(x_r)$

18.4 An iteration of the form $x_{r+1} = f(x_r)$; failure to give a specified root; an alternative iteration

18.5 Newton–Raphson method; comparison with $x_{r+1} = f(x_r)$

18.6 Newton–Raphson method; binomial expansion

18.1 Notes

Locating a root

If $y = f(x)$ is continuous between $x = a$ and $x = b$, there will be a root of the equation $f(x) = 0$ between $x = a$ and $x = b$ if and only if $f(a)$ and $f(b)$ differ in sign.

Newton–Raphson method

If $x = a$ is a good approximation to a root of $f(x) = 0$, a better approximation is often given by

$$x = a - \frac{f(a)}{f'(a)}$$

ie if x_r and x_{r+1} are successive approximations to $f(x) = 0$,

$$x_{r+1} = x_r - \frac{f(x_r)}{f'(x_r)}.$$

Other iterations

To find the root $x = a$ of an equation $f(x) = 0$ using an iteration

$$x_{r+1} = F(x_r),$$

this iteration will converge to the root $x = a$ if $|F'(a)| < 1$.

MATHEMATICS REVISION WORKBOOK

18.2 Locating a root; solving using Newton–Raphson method

Question

(a) Show there is a root of $x^3 - 5x + 1$ between 0 and 1.
(b) Find an approximation to this root, correct to 3 dp, using the Newton–Raphson method.

Hint Use our graphics calculator to graph the curve $y = x^3 - 5x + 1$.

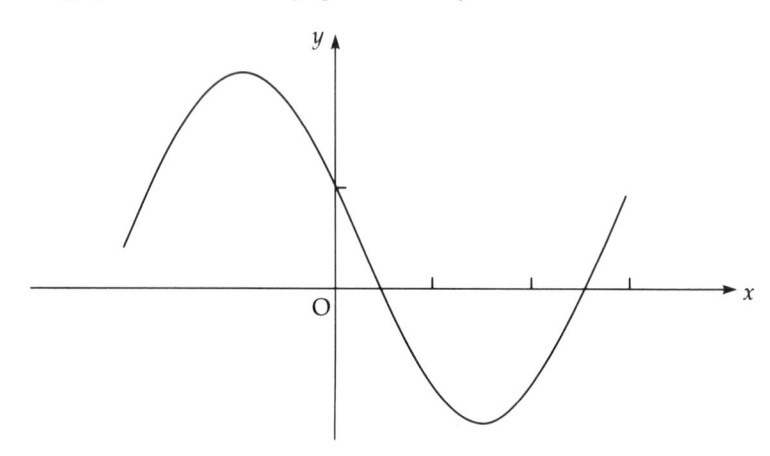

Solution

(a) Writing $f(x) = x^3 - 5x + 1$, $f(0) = 1$,
$$f(1) = -3.$$
As these differ in sign, and as $y = f(x)$ is continuous, there is a root of the equation $f(x) = 0$ between $x = 0$ and $x = 1$.

(b) The Newton–Raphson method says that if $x = a$ is a good approximation to a root of the equation $f(x) = 0$, then $x = a - \dfrac{f(a)}{f'(a)}$ is often a better approximation.
As $f(x) = x^3 - 5x + 1$, $f'(x) = 3x^2 - 5$.
Taking $x = 0$ as the first approximation, $f(0) = 1$, $f'(0) = -5$, so a better approximation is $0 - \dfrac{1}{-5}$, ie 0.2.
Now $f(0.2) = 0.008$, $f'(0.2) = -4.88$, so the next approximation is $0.2 - \dfrac{0.008}{(-4.88)}$,

about 0.201 639 ...
= 0.2016, to 4 dp.

Now $f(0.2016) = 0.000\ 194$ and $f'(0.2016) = -4.878$, so the next approximation is

136

ITERATIVE METHODS OF SOLVING EQUATIONS

$$0.2016 - \frac{0.000\,194}{(-4.878)} = 0.2016,$$

and, correct to 3 dp, the solution is

$$\underline{x = 0.202.}$$

Comments

Strictly, we ought to check that $f(x)$ changes sign between 0.2015 and 0.2025

$f(0.2015) = 0.000\,681$, positive,

$f(0.2025) = -0.04\,196$, negative,

so that the solution, correct to 3 dp, is $x = 0.202$.

All the calculations are shown in this solution but when we understand what is happening we can probably do all the calculations on our calculator without writing down any working. Calculators vary, but the following routine works on a Casio fx-7000GB:

0.2 EXE   Displays 0.2
Ans − (Ans³ − 5 × Ans + 1) ÷ (3 × Ans² − 5) EXE   Displays 0.201 639 344 3
                                              EXE   0.201 639 675 7
                                              EXE   0.201 639 675 7

showing that the root appears to be 0.201 639 676 to 9 dp.

18.3 An iteration of the form $x_{r+1} = f(x_r)$

Question

Use the iteration

$$x_{r+1} = e^{-x_r}$$

to find the solution near to $x = 0.6$ of

$$x = e^{-x},$$

giving your answer correct to 3 dp.

Solution

When $x_0 = 0.6$, $x_1 = e^{-0.6} = 0.54881\ldots$
When $x_1 = 0.5488$, $x_2 = 0.57764\ldots$
When $x_2 = 0.5776$, $x_3 = 0.56124\ldots$
When $x_3 = 0.5612$, $x_4 = 0.57052\ldots$
When $x_4 = 0.5705$, $x_5 = 0.56524\ldots$
When $x_5 = 0.5652$, $x_6 = 0.56824\ldots$
When $x_6 = 0.5682$, $x_7 = 0.56654\ldots$
When $x_7 = 0.5665$, $x_8 = 0.5675$
When $x_8, = 0.5675$, $x_9 = 0.56694\ldots$
When $x_9 = 0.5669$, $x_{10} = 0.56728\ldots$

These values are oscillating either side of the root, so that, correct to 3 dp, the solution is

$$\underline{x = 0.567.}$$

MATHEMATICS REVISION WORKBOOK

Comments

This can be done easily on many calculators without writing down much working. On the Casio fx-7000GB, the instructions are:

0.6	EXE	Display 0.6
e^(−Ans)	EXE	Display 0.5488116361
	EXE	Display 0.5776358443
	EXE	Display 0.5612236194
	EXE	Display 0.5705105488
	EXE	Display 0.5652367841 etc

For an examination, we have to show enough working for the examiner to follow what we are doing. The following would be acceptable.

Using the iteration
$$x_{r+1} = e^{-x_r}$$
when $x_0 = 0.6$, $x_1 = 0.5488$...... by calculator,
$x_2 = 0.57763$...... by calculator,

Later values in the iteration are
0.567255...
and 0.567079...

The values have been oscillating either side of the root, so that, correct to 3 dp, the solution is
$$x = 0.567.$$

Any iteration $x = f(x)$ converges to a root near $x = a$ if $|f'(a)| < 1$.
When $f(x) = e^{-x}$, $f'(x) = -e^{-x}$ and $f'(x) = e^{-0.6} = -0.548...$ so $|f'(0.6)| < 1$, and the iteration converges to this root.

This iteration is illustrated by the graphs of $y = x$ and $y = e^{-x}$ near the root; this is sometimes called a cobweb diagram.

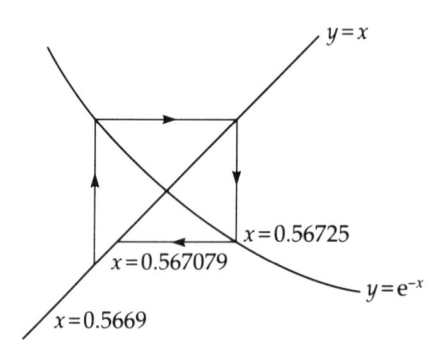

18.4 An iteration of the form $x_{r+1} = f(x_r)$; failure to give a specified root; an alternative iteration

Question

(a) Use the iteration
$$x_{r+1} = \tfrac{1}{6}(x_r^3 + 2)$$

ITERATIVE METHODS OF SOLVING EQUATIONS

to find the solution near to 0.4 of
$$x^3 - 6x + 2 = 0,$$
giving your solution to 3 dp.
(b) Show that the above iteration will not give the root that is near to 2.
(c) Find an iteration that will give the root near to 2, and use it to find this root correct to 3 sf.

Solution

(a) Taking $x_0 = 0.4$, we have $x_1 = \frac{1}{6}(0.4^3 + 2)$, ie 0.344,
When $x_1 = 0.344$, $x_2 = 0.340\ 12\ldots$
When $x_2 = 0.340\ 12$, $x_3 = 0.339\ 890\ldots$
Both these values correct to 0.340, to 3 sf, so
$$\underline{x = 0.340 \text{ is the solution, correct to 3 sf.}}$$ •

(b) An iteration $x_{r+1} = f(x_r)$ converges to a root near $x = a$ if and only if $|f'(a)| < 1$.
When $f(x) = \frac{1}{6}(x^3 + 2)$, $f'(x) = \frac{1}{2}x^2$,
so $\qquad f'(2) = 2$, which is greater than 1, so
$$\underline{\text{this iteration will not converge to the root near to 2.}}$$ ••

(c) We need to rearrange the equation $x^3 - 6x + 2 = 0$ into a form $x = f(x)$, where $|f'(2)| < 1$. When we try
$x = \sqrt[3]{(6x - 2)}$, ie $f(x) = (6x - 2)^{\frac{1}{3}}$
we have $\qquad f'(x) = \frac{1}{3}(6x - 2)^{-\frac{2}{3}}(6)$
and $\qquad f'(2) = \frac{1}{3}(10)^{-\frac{2}{3}}(6)$
$\qquad\qquad = 0.43\ldots$, which is less than 1
so since $|f'(2)| < 1$, this iteration converges to the root near to 2.
To find the root near to 2,
when $x_0 = 2$, $x_1 = \sqrt[3]{(10)}$, ie $2.154\ 4\ldots$
when $x_1 = 2.1544$, $x_2 = 2.219\ 0\ldots$
when $x_2 = 2.2190$, $x_3 = 2.244\ 9\ldots$
when $x_3 = 2.2449$, $x_4 = 2.255\ 1\ldots$
when $x_4 = 2.2552$, $x_5 = 2.259\ 2\ldots$
when $x_5 = 2.2592$, $x_6 = 2.260\ldots$
These are converging very slowly, but are approaching the value
$$\underline{2.26, \text{ to 3 sf.}}$$ •••

Comments

• We can probably do all these calculations on our calculator. The following set of instructions does this on a Casio fx-7000GB:

0.4	EXE	Display 0.4
(Ans³ + 2) ÷ 6	EXE	Display 0.344
	EXE	Display 0.340 117 9 …
	EXE	Display 0.339 890 8…

139

EXE   Display 0.339 877 6...
EXE   Display 0.339 876 9... etc

Strictly, before we conclude that the root is 0.340, to 3 dp, we should check that f (0.3395) and f (0.3405) differ in sign.

This convergence is illustrated by the diagram below, sometimes called a staircase diagram:

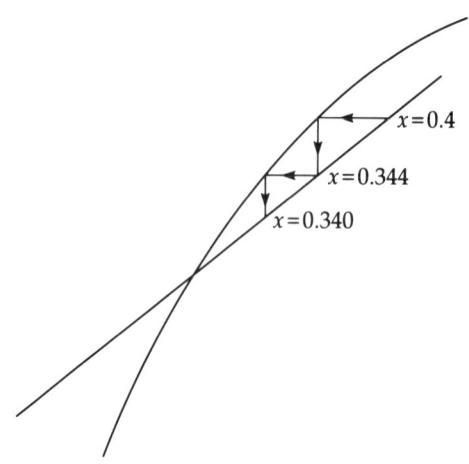

Using a calculator in this manner is so easy that it rarely matters whether an iteration converges very rapidly or not.

•• We can use the instructions at •, starting with
2 EXE
and obtain the sequence   1.666 666...
                           1.104 938...
                           0.558 167... etc
which shows that, even starting with the value $x_0 = 2$, this iteration converges to the root near to 0.340.

••• As with (a), using a calculator reduces the labour so much that there is little merit in spending a long time looking for an iteration that converges rapidly. The iteration

$$x_{r+1} = \sqrt{\left(6 - \frac{2}{x_r}\right)}$$

converges rapidly to the root ≈ 2.26, giving 2.261 802... but is less obvious than the iteration we used.

18.5 Newton–Raphson method; comparison with $x_{r+1} = f(x_r)$

Question

Show that the equation $x^3 - x^2 - 2 = 0$ has a root α which lies between 1 and 2.

(a) Using 1.5 as a first approximation for α, use the Newton–Raphson method once to obtain a second approximation for α, giving your answer to 3 decimal places.

ITERATIVE METHODS OF SOLVING EQUATIONS

(b) Show that the equation $x^3 - x^2 - 2 = 0$ can be arranged in the form $x = \sqrt[3]{[f(x)]}$ where $f(x)$ is a quadratic function
Use an iteration of the form $x_{n+1} = g(x_n)$ based on this rearrangement and with $x_1 = 1.5$ to find x_2 and x_3, giving your answers to 3 decimal places.
(AEB)

Solution

If $\quad f(x) = x^3 - x^2 - 2,$
$\quad\quad f(1) = -2, \ f(2) = 2,$

so $\quad\quad\quad\quad\quad$ there must be a root between 1 and 2.

(a) Since $f(x) = x^3 - x^2 - 2,$
$\quad\quad f'(x) = 3x^2 - 2x.$

By the Newton–Raphson method, if $x = 1.5$ is a good approximation to a root,

$$x = 1.5 - \frac{f(1.5)}{f'(1.5)}$$

is a better approximation.
But $f(1.5) = 1.5^3 - 1.5^2 - 2 = -0.875,$
and $f'(1.5) = 3(1.5)^2 - 2(1.5) = 3.75,$
so a better approximation than $x = 1.5$ is

$$x = 1.5 - \frac{(-0.875)}{3.75}$$

$\quad\quad = 1.733, \text{ to 3 decimal places.}$

(b) If $\quad x^3 - x^2 - 2 = 0,$
$\quad\quad x^3 = x^2 + 2,$
$\quad\quad\quad\quad x = \sqrt[3]{(x^2 + 2)}, \text{ which is in the form } x = \sqrt[3]{[f(x)]}.$

Taking $\quad x_1 = 1.5,$
$\quad\quad\quad x_2 = \sqrt[3]{(1.5^2 + 2)}$
$\quad\quad\quad\quad = 1.6198\ldots$
$\quad\quad\quad x_3 = 1.66596\ldots$

so $\quad\quad\quad\quad x_2 = 1.620, \ x_3 = 1.666, \text{ to 3 decimal places.}$

Comment

We should be able to use our calculator to work out these (and later) approximations very easily. Repeating the calculations shows that both the methods used here converge very slowly. The iteration

$$x_{r+1} = \sqrt{(x_r + 2/x_r)}$$

converges much more rapidly to the root $x = 1.695\,620\,77\ldots$

18.6 Newton–Raphson method; binomial expansion

Question

It is given that

$$f(x) = x - (\sin x + \cos x)^{\frac{1}{2}}, \quad 0 \leq x \leq \tfrac{3}{4}\pi.$$

(a) Show that the equation $f(x) = 0$ has a root lying between 1.1 and 1.2.
(b) Using 1.2 as a first approximation to this root, apply the Newton–Raphson procedure once to obtain a second approximation, giving your answer to 2 decimal places.
(c) When x is small enough for the terms in x^2 and higher powers of x to be neglected, use the approximations $\sin x \approx x$ and $\cos x \approx 1$ with the binomial expansion to show that

$$f(x) = A + Bx$$

and find the values of the constants A and B.

(L)

Hint Remember x is in RADIANS.

Solution

(a) Since $\quad f(1.1) = -0.0596\ldots\ldots$
and $\quad f(1.2) = 0.0622\ldots\ldots$
these differ in sign, and as $f(x)$ is continuous in this interval, there must be a root between $x = 1.1$ and $x = 1.2$.

(b) The Newton–Raphson formula is that if $x = a$ is a good approximation to a root of the equation $f(x) = 0$, then

$$x = a - \frac{f(a)}{f'(a)}$$

is often a better approximation.
Since $\quad f(x) = x - (\sin x + \cos x)^{\frac{1}{2}},$
$\quad f'(x) = 1 - \tfrac{1}{2}(\sin x + \cos x)^{-\frac{1}{2}}(\cos x - \sin x)$
so $\quad f(1.2) = 0.062\,284\ldots\ldots$
and $\quad f'(1.2) = 1.250\,36\ldots\ldots$
A better approximation to the root will be $1.2 - \dfrac{0.062\,284}{1.250\,36}$

$$= 1.150\,18$$
$$= \underline{\underline{1.15, \text{ to 2 dp.}}}$$

(c) When x is small,
$$f(x) = x - (1+x)^{\frac{1}{2}}, \text{ using } \sin x = x, \cos x = 1$$
$$= x - (1 + \tfrac{1}{2}x), \text{ neglecting terms in } x^2 \text{ and higher powers of } x,$$
so $f(x) \approx -1 + \tfrac{1}{2}x$

and $\underline{\underline{A = -1 \text{ and } B = \tfrac{1}{2}.}}$ •

Comment
- We may well feel glad to find an easy arithmetic check! When $x = 0.1$,
$$f(0.1) = -0.946\,344\ldots\ldots$$
and the approximation gives $-1 + \tfrac{1}{2}(0.1) = -0.95$.

19 COMPLEX NUMBERS

19.1 Notes
19.2 Modulus and argument of a complex number
19.3 Expressing a complex number in the form $a + ib$; modulus-argument form
19.4 Use of an Argand diagram; finding the argument of a complex number
19.5 Roots of a quadratic equation
19.6 Find all three roots of a cubic equation, given one complex root
19.7 Finding a locus

19.1 Notes

Complex conjugates

If $z = a + ib$, the complex conjugate of z, written z^* (sometimes $\bar{z}$) is $z^* = a - ib$.
If a polynomial with real coefficients has any complex roots, these roots occur in conjugate pairs, that is, **if z is a root, so is z^***.

Modulus and argument

The modulus of z, written $|z|$, is $\sqrt{(a^2 + b^2)}$.
The argument of z, written arg z, is invtan (b/a) *only if a is positive*. If a is negative, draw an Argand diagram as below:

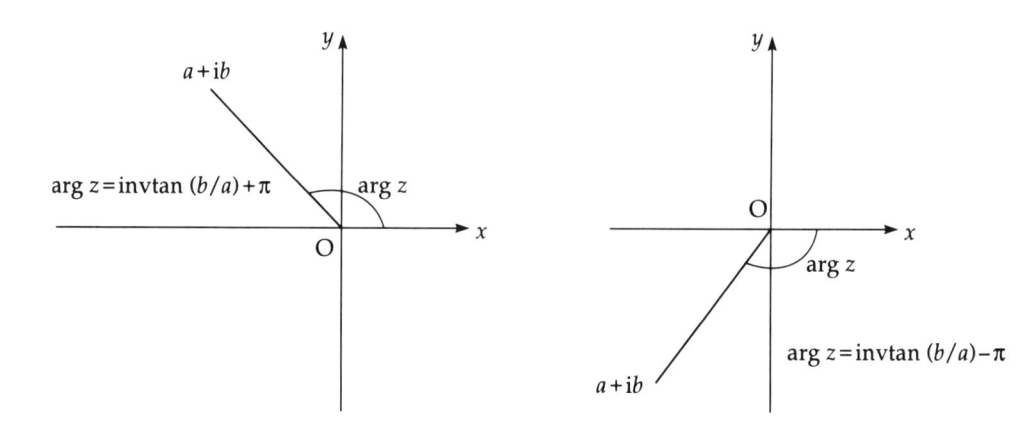

COMPLEX NUMBERS

Real and imaginary parts

To write the complex number $\dfrac{c+id}{a+ib}$ in the form showing its real and imaginary parts, multiply numerator and denominator by $a-ib$,

ie $\qquad \dfrac{(c+id)(a-ib)}{(a+ib)(a-ib)} = \dfrac{ac+bd+i(ad-bc)}{a^2+b^2}.$

Multiplication and division

$$r(\cos\alpha + i\sin\alpha) \times R(\cos\beta + i\sin\beta) = rR(\cos(\alpha+\beta) + i\sin(\alpha+\beta))$$

and $\qquad \dfrac{r(\cos\alpha+i\sin\alpha)}{R(\cos\beta+i\sin\beta)} = \dfrac{r}{R}(\cos(\alpha-\beta) + i\sin(\alpha-\beta)).$

19.2 Modulus and argument of a complex number

Question

Find the modulus and argument of the complex number $\dfrac{\sqrt{3}+i}{1+i\sqrt{3}}$.

(AEB)

Hint To write the number $\dfrac{a+ib}{c+id}$ in modulus-argument form, we can either multiply 'top' and 'bottom' by $c-id$, or we can express each of $a+ib$ and $c+id$ in modulus-argument form (see solution II).

Solution I

Write
$$\begin{aligned}
\dfrac{\sqrt{3}+i}{1+i\sqrt{3}} &= \dfrac{(\sqrt{3}+i)(1-i\sqrt{3})}{(1+i\sqrt{3})(1-i\sqrt{3})} \\
&= \dfrac{\sqrt{3}-3i+i+\sqrt{3}}{1+i\sqrt{3}-i\sqrt{3}+3} \\
&= \dfrac{2\sqrt{3}-2i}{4} \\
&= \tfrac{1}{2}(\sqrt{3}-i).
\end{aligned}$$

To write this number in modulus-argument form, it is often easiest to represent it on an Argand diagram, as below, where r is the modulus and θ is the argument. We see that $\underline{\underline{r=1}}$ and $\theta = \text{invsin}(-\tfrac{1}{2})$, ie $\underline{\underline{-\pi/6}}$.

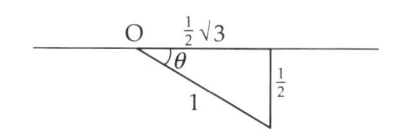

The number in modulus-argument form is
$$1(\cos(-\pi/6) + i\sin(-\pi/6)).$$

Solution II

Writing $\sqrt{3} + i$ and $1 + i\sqrt{3}$ each in modulus-argument form,

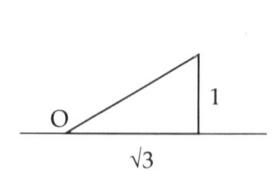

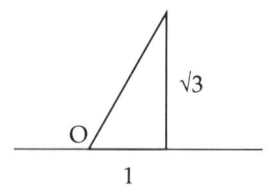

$$\sqrt{3} + i = 2(\cos(\pi/6) + i\sin(\pi/6))$$
and $\quad 1 + i\sqrt{3} = 2(\cos(\pi/3) + i\sin(\pi/3))$

so
$$\frac{\sqrt{3} + i}{1 + i\sqrt{3}} = \frac{2(\cos(\pi/6) + i\sin(\pi/6))}{2(\cos(\pi/3) + i\sin(\pi/3))}$$
$$= \cos(-\pi/6) + i\sin(-\pi/6), \text{ dividing the moduli and subtracting the arguments.}$$

Comment

- The complex number $a + ib$ is written $r(\cos\theta + i\sin\theta)$ in modulus-argument form, where $r = +\sqrt{(a^2 + b^2)}$ and $\cos\theta : \sin\theta : 1 = a : b : \sqrt{(a^2 + b^2)}$.
  Except for very simple values of θ, it is almost always best to draw an Argand diagram.

19.3 Expressing a complex number in the form $a + ib$; modulus-argument form

Question

The complex number $z = p + iq$, where p and q are real, satisfies the equation

$$\frac{2z + 16}{z + 5} = 3 - i.$$

Calculate

(i) the value of p and the value of q,
(ii) the argument of z giving your answer in radians, to two decimal places.

(AEB)

COMPLEX NUMBERS

Solution

(i) $\dfrac{2z + 16}{z + 5} = 3 - i$

$2z + 16 = (z + 5)(3 - i)$

$2p + 2iq + 16 = (p + iq + 5)(3 - i)$
$= 3p + 15 + q + i(3q - p - 5).$

Equating real parts, $2p + 16 = 3p + 15 + q$, ie $p + q = 1$ (1)
Equating imaginary parts, $2q = 3q - p - 5$, ie $p - q = -5$ (2)
<u>Adding (1) and (2), $p = -2$, $q = 3$.</u>

(ii) To find arg z, plot the point corresponding to $p + iq$ in an Argand diagram:

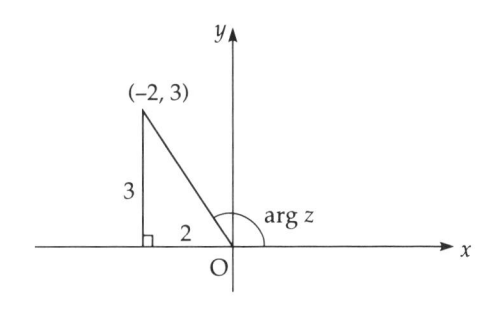

$$\arg z = \pi - \text{invtan}(3/2)$$
$$\underline{\underline{= 2.16 \text{ to 2 dp.}}}$$

19.4   Use of an Argand diagram; finding the argument of a complex number

Question

The complex numbers z and w are given by

$$z = \frac{A}{1 - i}, \quad w = \frac{B}{1 - 3i}$$

where A and B are real numbers. Given that $z + w = i$,

(a) show that $A = -1$ and find the value of B.

Using these values of A and B,

(b) show z and w on an Argand diagram,
(c) find $\arg(z - w)$.

147

MATHEMATICS REVISION WORKBOOK

Solution

(a) Since $$z = \frac{A}{1-i}$$
$$z = \frac{A(1+i)}{(1-i)(1+i)} = \frac{A+Ai}{2}.$$
Similarly $$w = \frac{B(1+3i)}{(1-3i)(1+3i)} = \frac{B+3Bi}{10}$$
so that $z + w = \frac{1}{2}(A + Ai) + \frac{1}{10}(B + 3Bi)$.

But $z + w = i$, so equating the real parts,

$\frac{1}{2}A + \frac{1}{10}B = 0$
and equating the imaginary parts,
$\frac{1}{2}A + \frac{3}{10}B = 1$.
Solving, $\qquad\qquad\qquad\underline{A = -1 \text{ and } B = 5}.$

(b) Using the form $z = \frac{1}{2}(A + Ai)$ and $w = \frac{1}{10}(B + 3Bi)$
$$z = -\tfrac{1}{2} - \tfrac{1}{2}i, \ w = \tfrac{1}{2} + \tfrac{3}{2}i$$
which are shown on the Argand diagram below:

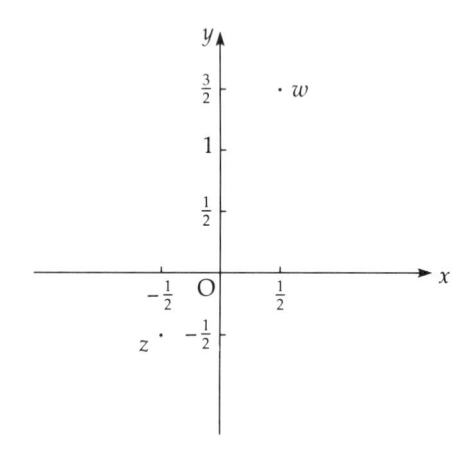

(c) $z - w = -\tfrac{1}{2} - \tfrac{1}{2}i - (\tfrac{1}{2} + \tfrac{3}{2}i)$
$= -1 - 2i$
so $\arg(z - w) = \text{invtan2} - \pi$ in radians, or $-116.6°$
as shown in the Argand diagram below:

148

COMPLEX NUMBERS

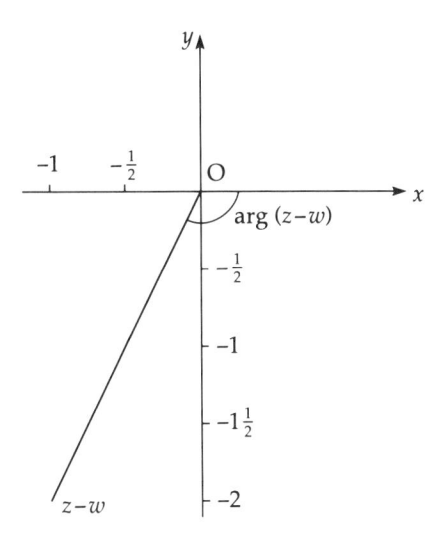

Notes

- Complex numbers are handled more easily if the denominator is real, eg $\dfrac{1+3i}{5}$, so multiply numerator and denominator by the complex conjugate of the denominator as here. If we try

$$\frac{A}{1-i} + \frac{B}{1-3i} = \frac{A(1-3i) + B(1+i)}{(1-i)(1-3i)}$$

  the arithmetic becomes very tricky.

- - Remember that if two complex numbers are equal, their real parts must be equal and their imaginary parts must be equal,

  ie $\qquad x + yi = A + Bi \Rightarrow x = A$ and $y = B$.

- - - When finding the argument of a complex number, it is vital to use an Argand diagram. The formula $\arg(x+yi) = \text{invtan}(y/x)$ is only true if x is positive, ie z is in the first or fourth quadrant in the Argand diagram.

149

MATHEMATICS REVISION WORKBOOK

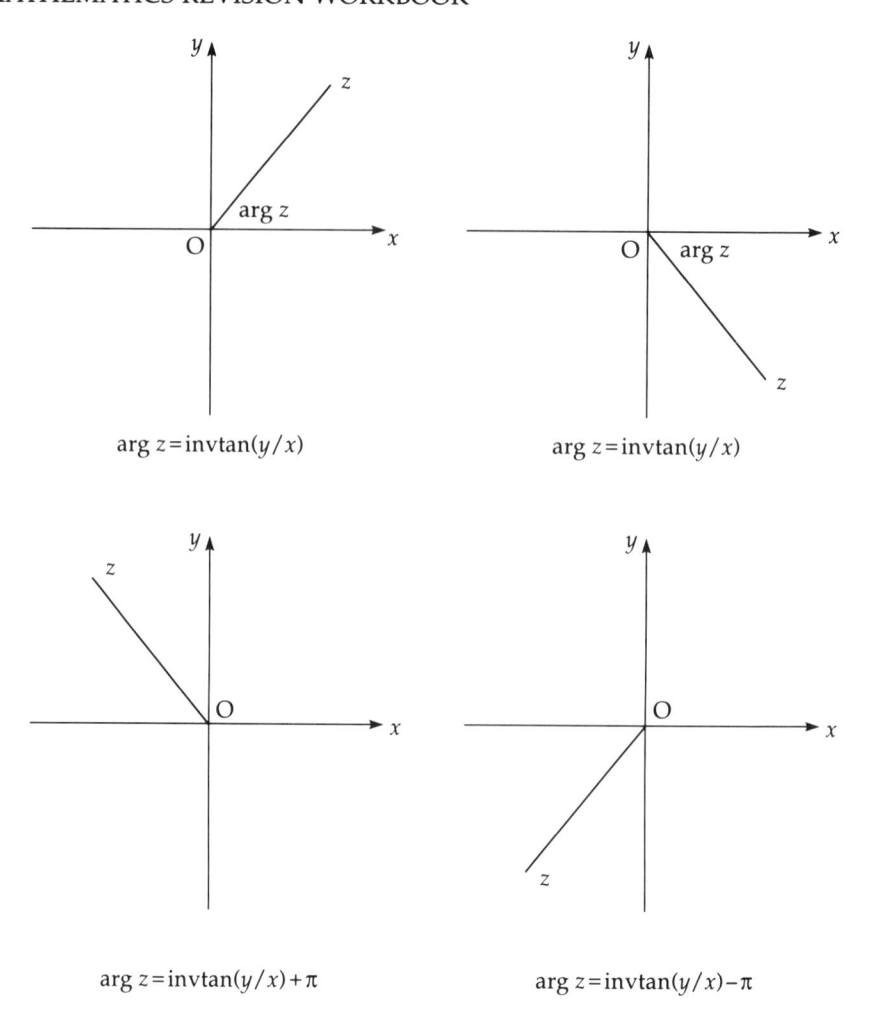

arg z = invtan(y/x)

arg z = invtan(y/x)

arg z = invtan(y/x) + π

arg z = invtan(y/x) − π

19.5 Roots of a quadratic equation

Question

(i) State the condition under which the quadratic equation $z^2 + pz + q = 0$, where p and q are real numbers, does not have real roots.

(ii) Show that $z_1 = -\frac{1}{2} + i\frac{\sqrt{3}}{2}$ is a root of the equation $z^2 + z + 1 = 0$, and find the other root z_2. Illustrate z_1 and z_2 on an Argand diagram.

Using the fact that z_1 satisfies the equation $z^2 + z + 1 = 0$,

150

COMPLEX NUMBERS

(iii) show that $z_1^3 = -(z_1^2 + z_1)$, and hence find the value of z_1^3,

(iv) calculate the value of $(1 + 2z_1 + 3z_1^2)(1 + 3z_1 + 2z_1^2)$.

(MEI)

Solution

(i) The quadratic $z^2 + pz + q = 0$ does not have real roots if $p^2 < 4q$.

(ii) Using the formula for the solution of a quadratic equation,

$$z^2 + z + 1 = 0, z = \frac{-1 \pm \sqrt{(1^2 - 4)}}{2}$$

$$z = -\tfrac{1}{2} + i\frac{\sqrt{3}}{2} \text{ or } -\tfrac{1}{2} - i\frac{\sqrt{3}}{2}.$$

The first of these expressions is z_1, so z_1 is a root of the equation.

The other must be z_2, so $z_2 = -\tfrac{1}{2} - i\frac{\sqrt{3}}{2}$.

They are illustrated below on an Argand diagram:

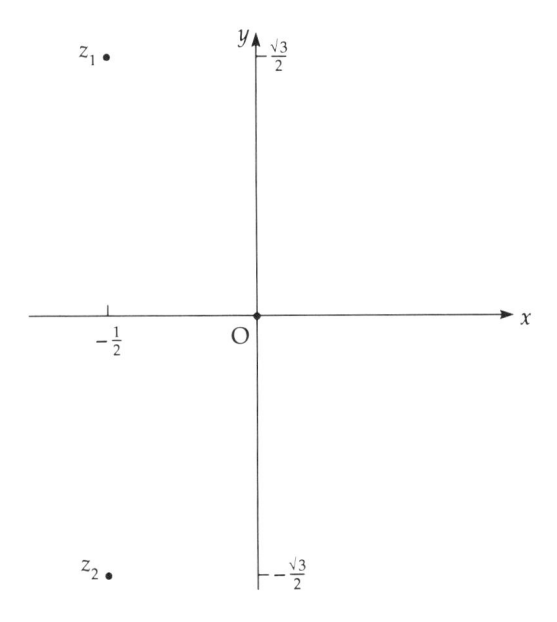

(iii) Since $z_1^2 + z_1 + 1 = 0$,

$z_1^3 + z_1^2 + z_1 = 0$,

so $z_1^3 = -(z_1^2 + z_1)$.

151

Since $\quad z_1^2 + z_1 + 1 = 0$
$\quad z_1^2 + z_1 = -1$
So $\quad z_1^3 = -(-1) = 1.$

(iv) Since $z_1^2 + z_1 + 1 = 0$, $1 + 2z_1 + 3z_1^2 = (1 + z_1 + z_1^2) + z_1 + 2z_1^2$
$$= z_1 + 2z_1^2.$$
Similarly, $\quad 1 + 3z_1 + 2z_1^2 = 2z_1 + z_1^2,$
so $(1 + 2z_1 + 3z_1^2)(1 + 3z_1 + 2z_1^2) = (z_1 + 2z_1^2)(2z_1 + z_1^2)$
$$= 2z_1^2 + 5z_1^3 + 2z_1^4$$
$$= 2z_1^2 + 5 + 2z_1, \text{ using } z_1^3 = 1,$$
$$= (2z_1^2 + 2z_1 + 2) + 3,$$
$$= 3, \text{ since } z_1^2 + z_1 + 1 = 0.$$

Notes

- The roots of a quadratic equation are $\dfrac{-b \pm \sqrt{(b^2 - 4ac)}}{2a}$, so the roots are real and distinct if $b^2 - 4ac > 0$,
  equal if $\quad b^2 - 4ac = 0$
  or not real if $\quad b^2 - 4ac < 0$.
- • It would be acceptable to verify that $z_1 = -\frac{1}{2} + i\frac{\sqrt{3}}{2}$ is a root of $z^2 + z + 1 = 0$ by substituting this value for z in the equation, and then finding the other root by noticing that the sum of the roots must be -1 (the sum of the roots of a quadratic is $-b/a$), so that as one root is $-\frac{1}{2} + i\frac{\sqrt{3}}{2}$, the other root must be $-\frac{1}{2} - i\frac{\sqrt{3}}{2}$, but this is longer than actually solving the equation.
- • • Since the question says 'Using the fact that z_1 satisfies the equation $z^2 + z + 1 = 0$', we should not do the later parts just by substituting $z_1 = -\frac{1}{2} + i\frac{\sqrt{3}}{2}$, though the examiners are likely to be sympathetic and award marks if we do. It is a long and tricky method to use!

19.6 Find all three roots of a cubic equation, given one complex root
Question
(i) Given that $f(z) = z^3 - 5z^2 + 8z - 6$, show that $1 - i$ is a root of $f(z) = 0$.
(ii) Find all the roots of $f(z) = 0$.

Hint We could substitute $z = 1 - i$ in $f(z)$, and show that $f(1 - i) = 0$, but since (ii) requires

COMPLEX NUMBERS

us to find all the roots of $f(z) = 0$, another approach will be quicker.
If $(a + ib)$ is a root of a polynomial equation with real coefficients, then $(a - ib)$ is also a root of that equation.

Solution

Since $z = 1 - i$ is a root of $f(z) = 0$, $z = 1 + i$ is also a root of that equation, and both $(z - 1 + i)$ and $(z - 1 - i)$ will be factors of $f(z)$.
Now $(z - 1 + i)(z - 1 - i) = z^2 - 2z + 2$, so $z^2 - 2z + 2$ must be a factor of $f(z)$,
ie
$$z^3 - 5z^2 + 8z - 6 = (z^2 - 2z + 2)(\ldots\ldots\ldots)$$
where the unknown factor on the right hand side is linear in z. Looking at the coefficient of z^3 on both sides, the unknown factor must have a term z; looking at the constant term on both sides, the unknown factor must have a term -3, so the unknown factor is $(z - 3)$, and

$$z^3 - 5z^2 + 8z - 6 = (z^2 - 2z + 2)(z - 3),$$
ie $\quad z^3 - 5z^2 + 8z - 6 = (z - 1 - i)(z - 1 + i)(z - 3),$
so $\quad \underline{\underline{1 - i \text{ is a root of } f(z) = 0.}}$

(ii) Since $f(z) = (z - 1 - i)(z - 1 + i)(z - 3)$,
the roots of $f(z) = 0$ are
$$\underline{\underline{1 - i, \ 1 + i \text{ and } 3.}}$$

19.7 Finding a locus

Question

In the Argand diagram, the point P represents the complex number z, where $z = x + iy$. Given that
$$z + 8 = \lambda \, i(z + 2),$$
where λ is real, equate real parts of this equation to show that $x + 8 = -\lambda y$. Find a second relation between x, y and λ, and eliminate λ to show that P always lies on a circle. Find the centre and radius of this circle.

(O&C)

Solution

Write z as $x + iy$, so $\quad z + 8 = \lambda \, i(z + 2)$
becomes $\quad x + iy + 8 = \lambda \, i(x + iy + 2)$
ie $\quad x + 8 + iy = -\lambda y + i \lambda (x + 2)$.

Equating the real parts of both sides,
$$x + 8 = -\lambda y,$$
equating the imaginary parts,

153

MATHEMATICS REVISION WORKBOOK

$$y = \lambda(x+2).$$

To eliminate λ between the equations, divide one by the other,

$$\frac{x+8}{y} = \frac{-y}{(x+2)}$$

ie $(x+8)(x+2) + y^2 = 0.$ \hfill (1)

This can be written

$$x^2 + 10x + 16 + y^2 = 0,$$

ie
$$x^2 + 10x + 25 + y^2 = 9,$$
$$(x+5)^2 + y^2 = 9,$$

which is the equation of a circle, centre $(-5, 0)$, radius 3.

Notes

From equation (1), we could see that this was an equation of the second degree in x and y, that the coefficient of x^2 was equal to that of y^2, and there was no term in xy, so that it must be a circle. As we were required to find the centre and the radius in this question, it was necessary to go on and put the circle in the form

$$(x-h)^2 + (y-k)^2 = r^2.$$

The number 25, of course, was found by completing the square of $x^2 + 10x$.

20 VECTORS

20.1 Notes
20.2 Components of equal vectors
20.3 Equations of two straight lines; angle between two lines
20.4 Displacement vectors and the angle between two vectors
20.5 Vectors written in matrix form; equations of two straight lines; angle between two straight lines
20.6 Perpendicular lines; image of a point
20.7 Equation of a plane; angle between a line and a plane
20.8 Equation of two straight lines; position vector of a point of intersection; angle between two straight lines; distance of a point from a line
20.9 Equation of a straight line; finding a line through a given point perpendicular to a given line; length of a line-segment

20.1 Notes

The magnitude of a vector

The magnitude of the vector $x\mathbf{i}+y\mathbf{j}$ is $\sqrt{(x^2+y^2)}$.
The magnitude of the vector $x\mathbf{i}+y\mathbf{j}+z\mathbf{k}$ is $\sqrt{(x^2+y^2+z^2)}$.

The scalar product

If $\mathbf{a}=a_1\mathbf{i}+a_2\mathbf{j}+a_3\mathbf{k}$ and $\mathbf{b}=b_1\mathbf{i}+b_2\mathbf{j}+b_3\mathbf{k}$, the scalar product of $\mathbf{a}$ and $\mathbf{b}$, written $\mathbf{a}.\mathbf{b}$, is

$$a_1b_1+a_2b_2+a_3b_3.$$

This follows from the definition of the scalar product

$$\mathbf{a}.\mathbf{b}=ab\ \cos\theta,$$

where θ is the angle between two vectors, so

$$\mathbf{i}.\mathbf{i}=\mathbf{j}.\mathbf{j}=\mathbf{k}.\mathbf{k}=1 \text{ and}$$

$$\mathbf{i}.\mathbf{j}=\mathbf{j}.\mathbf{k}=\mathbf{k}.\mathbf{i}=\mathbf{j}.\mathbf{i}=\mathbf{i}.\mathbf{k}=\mathbf{k}.\mathbf{j}=0.$$

MATHEMATICS REVISION WORKBOOK

Angle between two vectors

The angle θ between the vectors **a** and **b** is given by

$$\cos\theta = \frac{\mathbf{a}.\mathbf{b}}{ab},$$

where a, b are the magnitudes of **a** and **b** respectively. N.B. Work with $\mathbf{a}.\mathbf{b} = ab\cos\theta$, then divide by ab.

From this it follows that two vectors **a** and **b** are perpendicular if and only if

$$\mathbf{a}.\mathbf{b} = 0.$$

Equation of a straight line

The equation of the straight line through the point position vector **a** parallel to the vector **b** is given by

$$\mathbf{r} = \mathbf{a} + t\mathbf{b}.$$

This can be written in cartesian form

$$\frac{x - a_1}{b_1} = \frac{y - a_2}{b_2} = \frac{z - a_3}{b_3}.$$

Equation of a plane

The equation of the plane through the point position vector **a** perpendicular to the vector **n** is given by

$$(\mathbf{r} - \mathbf{a}).\mathbf{n} = 0$$

ie
$$\mathbf{r}.\mathbf{n} = \mathbf{a}.\mathbf{n}.$$

This can be written in cartesian form

$$a_1 x + a_2 y + a_3 z = a_1 n_1 + a_2 n_2 + a_3 n_3.$$

20.2 Components of equal vectors

Question

The vectors **a**, **b**, **c** are given by

$$\mathbf{a} = \mathbf{i} + 2\mathbf{j} + \mathbf{k}$$
$$\mathbf{b} = \mathbf{i} + \mathbf{j} + 2\mathbf{k}$$
$$\mathbf{c} = \mathbf{j} + \mathbf{k}.$$

A fourth vector **d** is given by

$$\mathbf{d} = 3\mathbf{i} + 6\mathbf{j} + 5\mathbf{k}.$$

Find values of α, β, γ such that

$$\mathbf{d} = \alpha\mathbf{a} + \beta\mathbf{b} + \gamma\mathbf{c}. \hspace{2cm} \text{(WJEC)}$$

Solution

Since $\mathbf{d} = \alpha\mathbf{a} + \beta\mathbf{b} + \gamma\mathbf{c}$
$3\mathbf{i} + 6\mathbf{j} + 5\mathbf{k} = \alpha(\mathbf{i} + 2\mathbf{j} + \mathbf{k}) + \beta(\mathbf{i} + \mathbf{j} + 2\mathbf{k}) + \gamma(\mathbf{j} + \mathbf{k}).$

Since the vectors **i**, **j** and **k** are mutually perpendicular, the **i**-components on both sides of the equation are equal, as are the **j**-components and the **k**-components, so

$$3 = \alpha + \beta \tag{1}$$
$$6 = 2\alpha + \beta + \gamma \tag{2}$$
$$5 = \alpha + 2\beta + \gamma. \tag{3}$$

Subtracting (3) from (2),

$$1 = \alpha - \beta. \tag{4}$$

Adding (1) and (4),

$$4 = 2\alpha,$$
$$\underline{\underline{\alpha = 2}}, \text{ whence } \underline{\underline{\beta = 1 \text{ and } \gamma = 1}}.$$

20.3 Equations of two straight lines; angle between two lines

Question

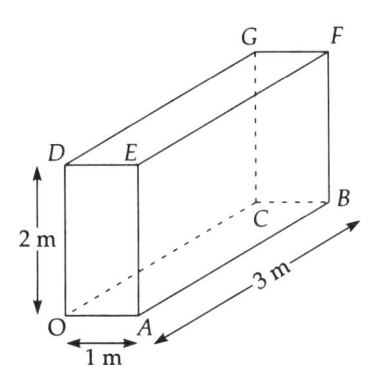

The figure shows a cuboid in which $OA = 1$ metre, $OC = 3$ metres and $OD = 2$ metres. Taking O as origin and unit vectors **i**, **j**, **k** in the directions OA, OC, OD respectively, express in terms of **i**, **j**, **k** the vectors

(i) **OF**
(ii) **AG**.

By considering an appropriate scalar product, find the acute angle between the diagonals OF and AG.

(WJEC)

MATHEMATICS REVISION WORKBOOK

Solution

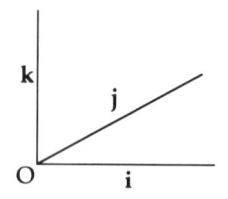

Using the unit vectors shown, the position vector OF of F relative to O is

$$i + 3j + 2k$$

as that displacement will take us from O to F.
Similarly the position vector AG of G relative to A is

$$-i + 3j + 2k.$$

The angle θ between two vectors **a**, **b** is given by the scalar product **a.b**, ie by

$$\mathbf{a.b} = ab \cos\theta$$

where $a = |\mathbf{a}|$, the magnitude of **a**, and $b = |\mathbf{b}|$ is the magnitude of **b**.

Here
$$|\mathbf{a}| = |i + 3j + 2k| = \sqrt{[1^2 + 3^3 + 2^2]}$$
$$= \sqrt{14}$$

and
$$|\mathbf{b}| = |-i + 3j + 2k| = \sqrt{[(-1)^2 + 3^3 + 2^2]}$$
$$= \sqrt{14},$$

so that

$$(i + 3j + 2k)_\bullet(-i + 3j + 2k); = \sqrt{14}.\sqrt{14} \cos\theta,$$

$$-1 + 9 + 4 = 14 \cos\theta,$$
$$\cos\theta = 6/7,$$
$$\underline{\underline{\theta = 31.0°, \text{ to 3 sf.}}}$$

20.4 Displacement vectors and the angle between two vectors

Question

Referred to an origin O, the points A and B have position vectors

$$5i + j + 2k \text{ and } -i + 7j + 8k.$$

The line l_1 passes through A and the line l_2 passes through B. The lines l_1 and l_2 intersect at the point C whose position vector is $i + 2j + k$.

VECTORS

(a) Find equations for the lines l_1 and l_2 giving each in the form $\mathbf{r} = \mathbf{a} + t\mathbf{b}$.
(b) Find the size of angle AOB, giving your answer to the nearest degree.

(L)

Hint Three-dimensional diagrams are difficult to draw, but a rough diagram showing the points O, A and B and the lines l_1 and l_2 will fix our ideas.

Solution

(a)

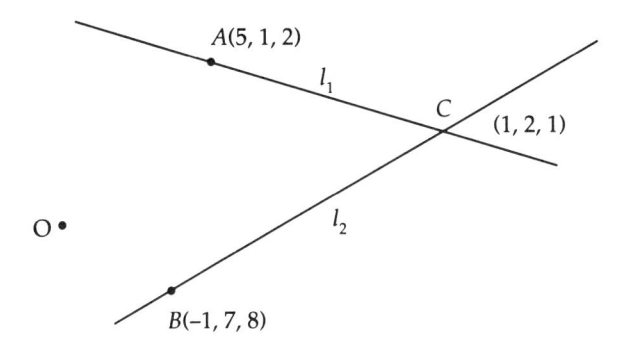

The line l_1 has direction-vector $\overrightarrow{AC}$, where
$$\overrightarrow{AC} = (\mathbf{i} + 2\mathbf{j} + \mathbf{k}) - (5\mathbf{i} + \mathbf{j} + 2\mathbf{k})$$
$$= -4\mathbf{i} + \mathbf{j} - \mathbf{k}.$$

Since l_1 goes through A, position vector $5\mathbf{i} + \mathbf{j} + 2\mathbf{k}$, its equation is
$$\underline{\underline{\mathbf{r} = (5\mathbf{i} + \mathbf{j} + 2\mathbf{k}) + t(-4\mathbf{i} + \mathbf{j} - \mathbf{k})}}.$$

Similarly, l_2 has direction-vector $\overrightarrow{BC}$, where
$$\overrightarrow{BC} = (\mathbf{i} + 2\mathbf{j} + \mathbf{k}) - (-\mathbf{i} + 7\mathbf{j} + 8\mathbf{k})$$
$$= 2\mathbf{i} - 5\mathbf{j} - 7\mathbf{k}$$

so that l_2 has equation
$$\mathbf{r} = (-\mathbf{i} + 7\mathbf{j} + 8\mathbf{k}) + t(2\mathbf{i} - 5\mathbf{j} - 7\mathbf{k}).$$

(b) The angle θ between the vectors **a** and **b** is given by
$$\mathbf{a}.\mathbf{b} = ab \cos\theta.$$

159

MATHEMATICS REVISION WORKBOOK

Here $\mathbf{a} = 5\mathbf{i} + \mathbf{j} + 2\mathbf{k}$, so $a \equiv |\mathbf{a}| = \sqrt{[5^2 + 1^2 + 2^2]}$
$= \sqrt{30}$

and $\mathbf{b} = -\mathbf{i} + 7\mathbf{j} + 8\mathbf{k}$, so $b \equiv |\mathbf{b}| = \sqrt{[(-1)^2 + 7^2 + 8^2]}$
$= \sqrt{114}$

and $(5\mathbf{i} + \mathbf{j} + 2\mathbf{k}).(-\mathbf{i} + 7\mathbf{j} + 8\mathbf{k}) = \sqrt{30}.\sqrt{114} \cos\theta$

ie $-5 + 7 + 16 = \sqrt{30}.\sqrt{114} \cos\theta$
$18 = \sqrt{30}.\sqrt{114} \cos\theta$,
$\cos\theta = \frac{18}{\sqrt{30}\sqrt{114}}$
$\theta = 72°$, to the nearest degree.

20.5 Vectors written in matrix form; equations of two straight lines; angle between two straight lines

Question

The lines l_1 and l_2 have equations

$$\mathbf{r} = \begin{pmatrix} 3 \\ 1 \\ 0 \end{pmatrix} + t \begin{pmatrix} 1 \\ 2 \\ 4 \end{pmatrix} \text{ and } \mathbf{r} = \begin{pmatrix} 1 \\ -1 \\ 1 \end{pmatrix} + s \begin{pmatrix} 2 \\ 1 \\ -1 \end{pmatrix}$$

respectively, where t and s are parameters. Show that l_1 passes through $(2, -1, -4)$ but that l_2 does not pass through this point.

Find the acute angle between l_2 and the line joining the points $(1, -1, 1)$ and $(2, -1, -4)$, giving your answer correct to the nearest degree.

(C)

Solution

If the line l_1 passes through $(2, -1, -4)$, we can find a value of t such that

$$\begin{pmatrix} 3 \\ 1 \\ 0 \end{pmatrix} + t \begin{pmatrix} 1 \\ 2 \\ 4 \end{pmatrix} = \begin{pmatrix} 2 \\ -1 \\ -4 \end{pmatrix}$$

ie such that
$3 + t = 2$
$1 + 2t = -1$
and $0 + 4t = -4.$

Obviously, the value $t = -1$ satisfies all three equations, so
l_1 passes through $(2, -1, -4)$.

To see if l_2 passes through $(2, -1, -4)$, we want a single value of s to satisfy all three equations

$$1 + 2s = 2,$$
$$-1 + s = -1,$$
$$1 - s = -4.$$

The first equation requires $s = \frac{1}{2}$, the second equation requires $s = 0$, so there is no value of s that satisfies all three equations, which shows that

$$\underline{\underline{l_2 \text{ does not pass through } (2, -1, -4).}}$$

The angle between two straight lines is found from their direction-vectors. The vector along the line joining $(1, -1, 1)$ and $(2, -1, -4)$ is

$$\begin{pmatrix} 2 \\ -1 \\ -4 \end{pmatrix} - \begin{pmatrix} 1 \\ -1 \\ 1 \end{pmatrix} \text{ ie } \begin{pmatrix} 1 \\ 0 \\ -5 \end{pmatrix}$$

so we require the angle between vectors **a** and **b**, where

$$\mathbf{a} = \begin{pmatrix} 1 \\ 0 \\ -5 \end{pmatrix} \text{ and } \mathbf{b} = \begin{pmatrix} 2 \\ 1 \\ -1 \end{pmatrix}.$$

Now $a \equiv |\mathbf{a}| = \sqrt{[1^2 + 0^2 + (-5)^2]}$, ie $\sqrt{26}$,
and $b \equiv |\mathbf{b}| = \sqrt{[2^2 + 1^2 + (-1)^2]}$, ie $\sqrt{6}$.

The angle θ between the two lines is given by $\mathbf{a}.\mathbf{b} = ab \cos\theta$
so

$$\begin{pmatrix} 1 \\ 0 \\ -5 \end{pmatrix} \cdot \begin{pmatrix} 2 \\ 1 \\ -1 \end{pmatrix} = \sqrt{26}.\sqrt{6} \cos\theta$$

ie
$$7 = \sqrt{26}.\sqrt{6} \cos\theta$$
$$\cos\theta = \frac{7}{\sqrt{26}.\sqrt{6}}$$
$$\underline{\underline{= 55.9°, \text{ to 3 sf.}}}$$

20.6 Perpendicular lines; image of a point

Question

A line l_1 passes through the point A, with position vector $5\mathbf{i} + 3\mathbf{j}$, and the point B, with position vector $-2\mathbf{i} - 4\mathbf{j} + 7\mathbf{k}$.

(a) Write down an equation of the line l_1.

A second line l_2 has equation
$$\mathbf{r} = \mathbf{i} - 3\mathbf{j} - 4\mathbf{k} + \mu(\mathbf{i} + 2\mathbf{j} + 3\mathbf{k})$$
where μ is a parameter.

(b) Show that l_1 and l_2 are perpendicular to each other.
(c) Show that the two lines meet, and find the position vector of the point of intersection

The point C has position vector $2\mathbf{i} - \mathbf{j} - \mathbf{k}$.

(d) Show that C lies on l_2.

The point D is the image of C after reflection in the line l_1.

(e) Find the position vector of D.

(L)

Solution

(a)

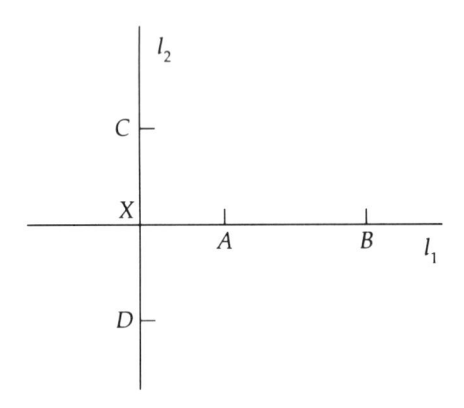

The vector AB is $(-2\mathbf{i} - 4\mathbf{j} + 7\mathbf{k}) - (5\mathbf{i} + 3\mathbf{j}) = -7\mathbf{i} - 7\mathbf{j} + 7\mathbf{k}$, so the equation of the line l_1 through A, position vector $(5\mathbf{i} + 3\mathbf{j})$ along AB is
$$\underline{\mathbf{r} = 5\mathbf{i} + 3\mathbf{j} + \lambda(-7\mathbf{i} - 7\mathbf{j} + 7\mathbf{k}).}$$

(b) Two vectors are perpendicular if their scalar product is zero, so l_1 and l_2 are perpendicular if the scalar product of their direct-vectors is zero. Here
$$(-7\mathbf{i} - 7\mathbf{j} + 7\mathbf{k}) \cdot (\mathbf{i} + 2\mathbf{j} + 3\mathbf{k}) = -7 - 14 + 21$$
$$= 0,$$
so the lines l_1 and l_2 are perpendicular.

(c) To show that the two lines meet,
$$\mathbf{r} = 5\mathbf{i} + 3\mathbf{j} + \lambda(-7\mathbf{i} - 7\mathbf{j} + 7\mathbf{k})$$
and $\mathbf{r} = \mathbf{i} - 3\mathbf{j} - 4\mathbf{k} + \mu(\mathbf{i} + 2\mathbf{j} + 3\mathbf{k})$

must have a common point,

ie
$$5 - 7\lambda = 1 + \mu \quad (1)$$
$$3 - 7\lambda = -3 + 2\mu \quad (2)$$
$$7\lambda = -4 + 3\mu \quad (3)$$

must have a common solution in λ and μ.
Subtracting (2) from (1),
$$2 = 4 - \mu, \quad \text{ie } \mu = 2.$$

Substituting in (2)
$$7\lambda = 2$$

and these values satisfy (3), so that the two lines meet at the point position vector $\mathbf{i} - 3\mathbf{j} - 4\mathbf{k} + 2(\mathbf{i} + 2\mathbf{j} + 3\mathbf{k})$
ie
$$\underline{\underline{3\mathbf{i} + \mathbf{j} + 2\mathbf{k}.}}$$

(d) To show that C, position vector $2\mathbf{i} - \mathbf{j} - \mathbf{k}$, lies on l_2, there must be a value for μ, so that
$$\mathbf{i} - 3\mathbf{j} - 4\mathbf{k} + \mu(\mathbf{i} + 2\mathbf{j} + 3\mathbf{k}) = 2\mathbf{i} - \mathbf{j} - \mathbf{k}.$$

This is obviously satisfied by $\mu = 1$, so
$$\underline{\underline{\text{the point } C \text{ lies on } l_2.}}$$

(e) Looking at the diagram, as D is the image of C in l_1, $CX = XD$, where X is the point of intersection of l_1 and l_2. Since
$$C \text{ has position vector } 2\mathbf{i} - \mathbf{j} - \mathbf{k},$$
$$\text{and } X \text{ has position vector } 3\mathbf{i} + \mathbf{j} + 2\mathbf{k},$$
$$\underline{\underline{D \text{ has position vector } 4\mathbf{i} + 3\mathbf{j} + 5\mathbf{k},}}$$

so that vectors $\overrightarrow{CX}$ and $\overrightarrow{XD}$ are equal.

20.7 Equation of a plane; angle between a line and a plane

Question

With respect to an origin, O, the straight lines l_1 and l_2 have equations

$$l_1 : \mathbf{r} = p\mathbf{i} - 2\mathbf{j} + 2\mathbf{k} + \lambda(\mathbf{i} - \mathbf{k})$$
$$l_2 : \mathbf{r} = 3\mathbf{i} - \mathbf{j} + \mu(2\mathbf{i} + \mathbf{j} - 3\mathbf{k}),$$

where λ and μ are parameters and p is a scalar constant. The lines intersect at the point A.

(a) Find the coordinates of A and show that $p = 2$.

The plane Π passes through A and is perpendicular to l_2.

(b) Find a cartesian equation of Π.

(c) Find the acute angle between the plane Π and the line l_1, giving your answer in degrees to 1 decimal place.

(L)

Solution

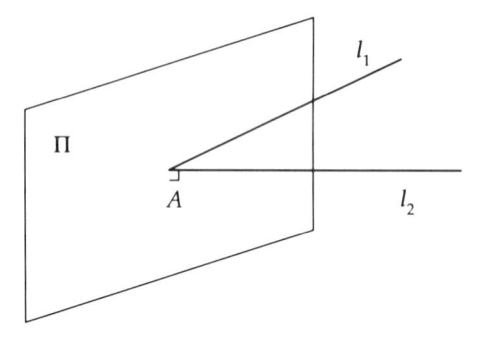

At A, the point of intersection of l_1 and l_2, the **i**-components must be equal, the **j**-components must be equal, as must the **k**-components, since

on l_1, $\quad \mathbf{r} = p\mathbf{i} - 2\mathbf{j} + 2\mathbf{k} + \lambda(\mathbf{i} - \mathbf{k})$, and

on l_2, $\quad \mathbf{r} = 3\mathbf{i} - \mathbf{j} + \mu(2\mathbf{i} + \mathbf{j} - 3\mathbf{k})$,

$$p + \lambda = 3 + 2\mu \qquad (1)$$
$$-2 = -1 + \mu \qquad (2)$$
$$2 - \lambda = -3\mu \qquad (3)$$

From (2) $\quad \mu = -1$

Putting $\mu = -1$ in (3), $\lambda = -1$

Putting $\mu = -1$ and $\lambda = -1$ in (1), $p - 1 = 3 + 2(-1)$,

$$\underline{\underline{p = 2.}}$$

(b) To find the position vector **a** of A, put $\mu = -1$ in l_2,

$$\mathbf{a} = 3\mathbf{i} - \mathbf{j} + (-1)(2\mathbf{i} + \mathbf{j} - 3\mathbf{k})$$
$$= \mathbf{i} - 2\mathbf{j} + 3\mathbf{k}.$$

The equation of a plane through a point position vector **a** perpendicular to a vector **n** is

$$\mathbf{r} \cdot \mathbf{n} = \mathbf{a} \cdot \mathbf{n} \quad (\text{or } \mathbf{r} \cdot (\mathbf{a} - \mathbf{n}) = 0)$$

Here, **n** is the direction-vector of l_2, so the equation of Π is

$$\mathbf{r} \cdot (2\mathbf{i} + \mathbf{j} - 3\mathbf{k}) = (\mathbf{i} - 2\mathbf{j} + 3\mathbf{k}) \cdot (2\mathbf{i} + \mathbf{j} - 3\mathbf{k})$$
$$(x\mathbf{i} + y\mathbf{j} + z\mathbf{k}) \cdot (2\mathbf{i} + \mathbf{j} - 3\mathbf{k}) = 2 - 2 - 9$$
$$\underline{\underline{2x + y - 3z + 9 = 0.}}$$

(c) The angle between the plane Π and the line l_1 is the complement of the angle between the lines l_1 and l_2. These lines have direction-vectors

$$\mathbf{i} - \mathbf{k} \text{ and } 2\mathbf{i} + \mathbf{j} - 3\mathbf{k}$$

so the angle θ between the two lines is given by

$$|\mathbf{i} - \mathbf{k}| \cdot |2\mathbf{i} + \mathbf{j} - 3\mathbf{k}| \cos\theta = (\mathbf{i} - \mathbf{k}) \cdot (2\mathbf{i} + \mathbf{j} - 3\mathbf{k})$$

ie $\quad\quad\quad\quad \sqrt{2}\sqrt{14} \cos\theta = 5,$

$$\cos\theta = \frac{5}{\sqrt{2}\sqrt{14}}$$
$$\theta = 19.1°,$$

the angle between l_1 and the plane $\underline{\underline{\Pi \text{ is } 70.9°}}$.

20.8 Equation of two straight lines; position vector of a point of intersection; angle between two straight lines; distance of a point from a line

Question

The points P and Q have position vectors $\mathbf{p} = 3\mathbf{i} - \mathbf{j} + 2\mathbf{k}$, $\mathbf{q} = 4\mathbf{i} - 2\mathbf{j} - \mathbf{k}$ respectively, relative to a fixed origin O.

(a) Determine a vector equation of the line l_1, passing through P and Q in the form $\mathbf{r} = \mathbf{a} + s\mathbf{b}$, where s is a scalar parameter.
(b) The line l_2 has vector equation $\mathbf{r} = 2\mathbf{i} - 2\mathbf{j} - 3\mathbf{k} + t(2\mathbf{i} - \mathbf{j} - 2\mathbf{k})$. Show that l_1 and l_2 intersect and find the position vector of the point of intersection V.
(c) Show that PV has length $3\sqrt{11}$.
(d) The acute angle between l_1 and l_2 is θ. Show that $\theta = \frac{3}{\sqrt{11}}$.
(e) Calculate the perpendicular distance from P to l_2. (AEB)

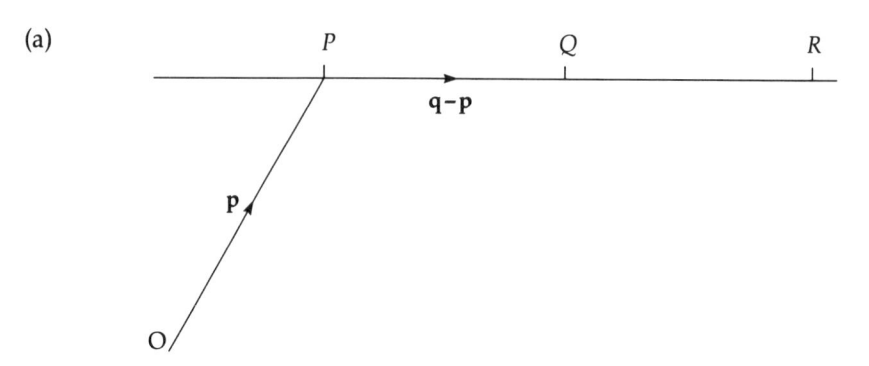

(a)

```
           P         Q              R
    ───────┼─────────┼──────────────┼──
           │   q-p
         P │
           │
         O │
```

The vector $\vec{PQ} = \mathbf{q} - \mathbf{p}$
$$= (4\mathbf{i} - 2\mathbf{j} - \mathbf{k}) - (3\mathbf{i} - \mathbf{j} + 2\mathbf{k})$$
$$= \mathbf{i} - \mathbf{j} - 3\mathbf{k}.$$

For any point R on the line PQ,
$$\vec{OR} = \vec{OP} + s\,\vec{PQ},$$
so if $\mathbf{r}$ is the position vector of R,
$$\underline{\mathbf{r} = 3\mathbf{i} - \mathbf{j} + 2\mathbf{k} + s(\mathbf{i} - \mathbf{j} - 3\mathbf{k}).}$$

(b) The lines l_1 and l_2 meet if there are values of s and t for which
$$3\mathbf{i} - \mathbf{j} + 2\mathbf{k} + s(\mathbf{i} - \mathbf{j} - 3\mathbf{k}) = 2\mathbf{i} - 2\mathbf{j} - 3\mathbf{k} + t(2\mathbf{i} - \mathbf{j} - 2\mathbf{k}).$$

These expressions are vectors, and so will only be equal if the $\mathbf{i}$-components are equal, the $\mathbf{j}$-components are equal and the $\mathbf{k}$-components are equal,

ie $3 + s = 2 + 2t, \quad \Rightarrow 1 = 2t - s$ (1)

$-1 - s = -2 - t, \quad \Rightarrow 1 = -t + s$ (2)

$2 - 3s = -3 - 2t, \quad \Rightarrow 5 = 3s - 2t.$ (3)

Adding (1) and (2), $t = 2$, $s = 3$.

These values satisfy (3), so the lines l_1 and l_2 do meet, and the point of intersection V is found by putting $s = 3$ in l_1 (or $t = 2$ in l_2). If V has position vector $\mathbf{v}$,
$$\underline{\mathbf{v} = 6\mathbf{i} - 4\mathbf{j} - 7\mathbf{k}.}$$

(c) The length PV
$$\begin{aligned} &= |\overrightarrow{PV}| = |\mathbf{v} - \mathbf{p}| \\ &= |(3\mathbf{i} - \mathbf{j} + 2\mathbf{k}) - (6\mathbf{i} - 4\mathbf{j} - 7\mathbf{k})| \\ &= |-3\mathbf{i} + 3\mathbf{j} + 9\mathbf{k}| \\ &= \sqrt{[(-3)^2 + (3)^2 + (9)^2]} \\ &= \sqrt{99} \\ &= 3\sqrt{11}. \end{aligned}$$

(d) The direction of the line $\mathbf{r} = \mathbf{a} + s\,\mathbf{b}$ is given by the vector $\mathbf{b}$, so the direction of the line l_1 is along $\mathbf{i} - \mathbf{j} - 3\mathbf{k}$, of l_2 is along $2\mathbf{i} - \mathbf{j} - 2\mathbf{k}$.
Use the scalar product to find the angle θ between these two vectors,

ie $\quad (\mathbf{i} - \mathbf{j} - 3\mathbf{k}) \cdot (2\mathbf{i} - \mathbf{j} - 2\mathbf{k}) = |\mathbf{i} - \mathbf{j} - 3\mathbf{k}| \cdot |2\mathbf{i} - \mathbf{j} - 2\mathbf{k}| \cos\theta$

ie $\quad\quad\quad\quad\quad\quad\quad 9 = \sqrt{11} \times 3 \ \cos\theta,$ **∙∙**

$$\cos\theta = \underline{\underline{\frac{3}{\sqrt{11}}}}.$$

(e)

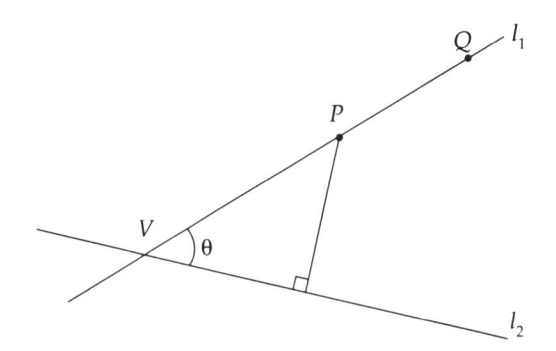

The perpendicular distance of P from l_2 is $PV \sin\theta$.

Since $\quad\quad\quad\quad\quad \cos\theta = \dfrac{3}{\sqrt{11}}, \ \sin\theta = \dfrac{\sqrt{2}}{\sqrt{11}}$ **∙∙∙**

so $\quad\quad\quad\quad\quad PV \sin\theta = 3\sqrt{11} \times \dfrac{\sqrt{2}}{\sqrt{11}}$

$\quad\quad\quad\quad\quad\quad\quad\quad\quad = \underline{\underline{3\sqrt{2}.}}$

Notes

● Each value of the parameter s determines a different point on the line PQ. The value $s = 0$ gives the point P, the value $s = 1$ gives Q. The value $s = \tfrac{1}{2}$ gives the mid-point of PQ. Values greater than $s = 1$ give points along PQ beyond Q; negative values give points along QP beyond P. There is a one-to-one correspondence between values of s and points on the line PQ. Any point on PQ can be taken in this form, for a suitable value of s.

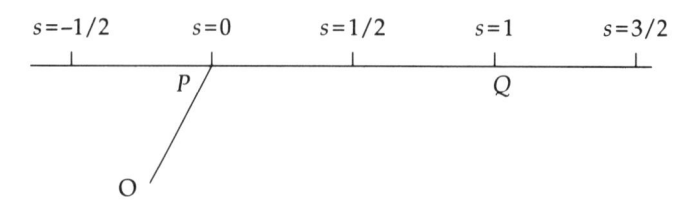

- The scalar product of the vectors a,b, written **a.b**, is defined by $\mathbf{a}.\mathbf{b} = ab \cos\theta$, where θ is one of the angles between the vectors. Thus $\mathbf{i}.\mathbf{i} = \mathbf{j}.\mathbf{j} = \mathbf{k}.\mathbf{k} = 1$ and $\mathbf{i}.\mathbf{j} = \mathbf{j}.\mathbf{k} = \mathbf{k}.\mathbf{i} = 0$, so

$$(a\mathbf{i} + b\mathbf{j} + c\mathbf{k}).(x\mathbf{i} + y\mathbf{j} + z\mathbf{k}) = ax + by + ck;$$

here $(\mathbf{i} - \mathbf{j} - 3\mathbf{k}).(2\mathbf{i} - \mathbf{j} - 2\mathbf{k}) = 1 \times 2 + (-1) \times (-1) + (-3) \times (-2)$

$$= 2 + 1 + 6$$
$$= 9.$$

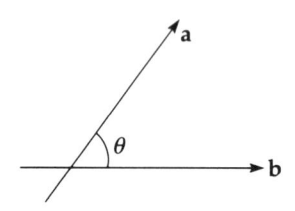

If we want to ensure that we have the acute angle between two vectors meeting at a point P, we have to use $\overrightarrow{PQ}.\overrightarrow{PR}$, not $\overrightarrow{PQ}.\overrightarrow{RP}$ or $\overrightarrow{QP}.\overrightarrow{PR}$:

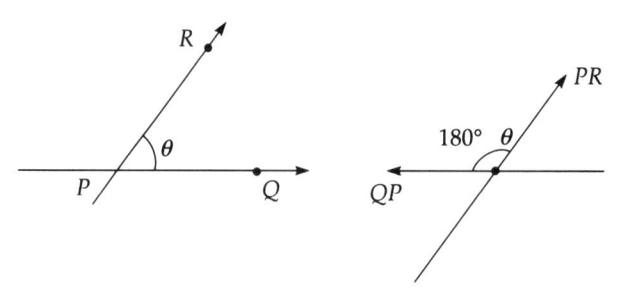

$\overrightarrow{PQ}.\overrightarrow{PR} = |PQ|\,|PR|\cos\theta$ $\overrightarrow{QP}.\overrightarrow{PR} = |QP|\,|PR|\cos(180° - \theta)$

- - - Given one trig ratio, here $\cos\theta$, it is usually best to draw a right-angled triangle to find any other ratios. Here

$$\cos\theta = \frac{3}{\sqrt{11}} \Rightarrow \sin\theta = \frac{\sqrt{2}}{\sqrt{11}} \text{ and } \tan\theta = \frac{\sqrt{2}}{3}.$$

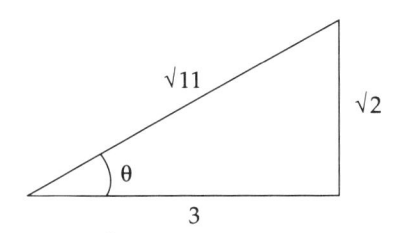

If we want the perpendicular distance of a point from a plane, we use the vector normal to the plane below (see Q20.9). The perpendicular distance of a point from a line is best found using trigonometry as in this solution.

20.9 Equation of a straight line; finding a line through a given point perpendicular to a given line; length of a line-segment

Question

In bad weather, the roof of a barn begins to sag. It is decided to support it as shown in the picture.

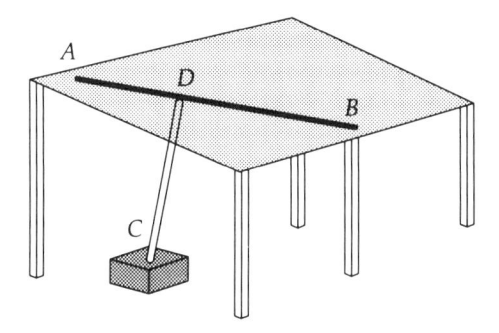

When supported, ADB is a straight line. Two points on the roof are $A(2, 0, 15)$ and $B(14, 9, 9)$, relative to an arbitrary origin.

(i) Find the equation of the line AB in vector form.

The support CD, resting on concrete blocks at C, is perpendicular to the line AB. C is the point $(3, -1, 1)$.

(ii) Write down the vector CP, where P is a general point on the line AB. Hence, using a scalar product, find the co-ordinates of D on AB such that CD is perpendicular to AB.

    (iii) Calculate the length of the support CD.
    (iv) Calculate the ratio AD:DB.

(MEI)

Solution

(i) Since A has coordinates $(2, 0, 15)$ and B has coordinates $(14, 9, 9)$, the vector $\overrightarrow{AB}$ is $12\mathbf{i} + 9\mathbf{j} - 6\mathbf{k}$. Thus the equation of the line AB is

$$\mathbf{r} = (2\mathbf{i} + 15\mathbf{k}) + t(12\mathbf{i} + 9\mathbf{j} - 6\mathbf{k}).$$

(ii) If $\mathbf{p}$ is the position vector of P, a general point on the line AB,

$$\mathbf{p} = (2\mathbf{i} + 15\mathbf{k}) + t(12\mathbf{i} + 9\mathbf{j} - 6\mathbf{k}), \text{ for varying values of } t.$$

If $\mathbf{c}$ is the position vector of C, the vector $\overrightarrow{CP} = \mathbf{p} - \mathbf{c}$, where

$$\mathbf{p} - \mathbf{c} = (2\mathbf{i} + 15\mathbf{k}) + t(12\mathbf{i} + 9\mathbf{j} - 6\mathbf{k}) - (3\mathbf{i} - \mathbf{j} + \mathbf{k})$$
$$= (-1 + 12t)\mathbf{i} + (9t + 1)\mathbf{j} + (14 - 6t)\mathbf{k}.$$

As $\overrightarrow{CP}$ is perpendicular to $\overrightarrow{AB}$, $\overrightarrow{CP} \cdot \overrightarrow{AB} = 0$

ie $[(-1 + 12t)\mathbf{i} + (9t + 1)\mathbf{j} + (14 - 6t)\mathbf{k}] \cdot [12\mathbf{i} + 9\mathbf{j} - 6\mathbf{k}]$

ie $12(-1 + 12t) + 9(9t + 1) - 6(14 - 6t) = 0$

ie $261t - 87 = 0,$

$$t = \frac{1}{3}$$

and the coordinates of D are found by putting $t = \frac{1}{3}$ in the expression for a general point on AB,

$$\mathbf{r} = (2\mathbf{i} + 15\mathbf{k}) + t(12\mathbf{i} + 9\mathbf{j} - 6\mathbf{k})$$

ie $6\mathbf{i} + 3\mathbf{j} + 13\mathbf{k}.$

(iii) The length of the support CD is the magnitude of the vector $\mathbf{p} - \mathbf{c}$ when $t = \frac{1}{3}$, so when $t = \frac{1}{3}$,

$$\mathbf{p} - \mathbf{c} = (-1 + 12t)\mathbf{i} + (9t + 1)\mathbf{j} + (14 - 6t)\mathbf{k} \text{ with } t = \tfrac{1}{3}.$$
$$= 3\mathbf{i} + 4\mathbf{j} + 12\mathbf{k},$$

ie $|\mathbf{p} - \mathbf{c}| = \sqrt{(3^3 + 4^2 + 12^2)}$
                   $= 13,$

the length of the support is 13 units.

VECTORS

(iv) The parameter t defines a point along AB. When $t = 0$, we have the point A. When $t = 1$ we have the point B. When $t = \frac{1}{3}$ we have the point one-third of the way along AB, so
$$\underline{AD : DB = 1 : 2}.$$

Notes

1. Perpendicular lines or vectors should always suggest the scalar product; here, of course, we had a hint to use it.
2. In vector questions at A-level, the numbers are usually chosen so that the final answers are not too awkward, so when we found $t = \frac{1}{3}$, we could feel confident that we had not made an arithmetic error.

21 PROBABILITY

21.1 Notes
21.2 Independent events. Given $P(A)$, $P(A \cup B)$, find $P(B)$, etc
21.3 Dependent events. Given $P(A)$, $P(B)$, $P(B \mid A)$, find $P(B' \mid A)$, etc
21.4 Use of Venn diagrams

21.1 Notes

Independent events

Two events A and B are independent if 'it does not matter whether one, say event B, has occurred or not', ie
$$P(A) = P(A \mid B) = P(A \mid B'),$$
whence $$P(A \mid B) = \frac{P(A \cap B)}{P(B)}$$
and $$P(B \mid A) = \frac{P(A \cap B)}{P(A)}.$$

When A and B are independent,
$$P(A \cap B) = P(A).P(B)$$

These are most easily illustrated by a Venn diagram. If, x, y, z and w are the number of elements in each of the sets illustrated, then

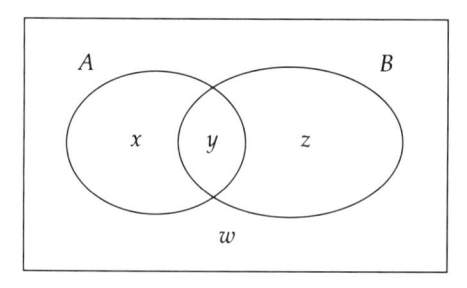

$$P(A) = \frac{P(A \cap B)}{P(B)} \Rightarrow \frac{x+y}{x+y+z+w} = \frac{y}{y+z}$$
$$= \frac{x}{x+w}, \text{ by algebra}$$

$$\Rightarrow P(A) = \frac{P(A \mid B')}{P(B')}, \text{ etc.}$$

172

PROBABILITY

21.2 Independent events. Given P(A), P(A ∪ B), find P(B), etc

Question

The events A, B are independent. Given that $P(A) = 0.2$ and $P(A \cup B) = 0.5$, calculate

(a) $P(B)$,
(b) $P(A' \cap B')$,
(c) $P(A \mid A \cup B)$.

(WJEC)

Hint These questions are made much clearer when we have drawn one – or more – Venn diagrams. The data are illustrated below:

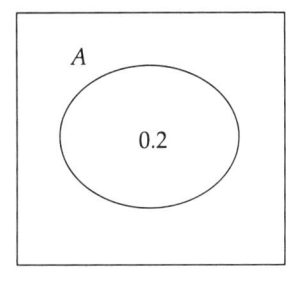

$P(A) = 0.2$ 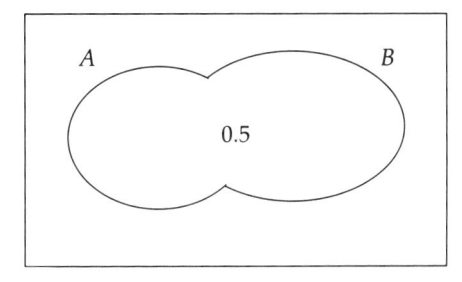　　　　　　　　　　$P(A \cup B) = 0.5$

We also have to note that we are given A, B are independent, so that $P(A \mid B) = P(A)$. This means that the probability of event A is the same whether A is chosen from B or not. When A and B are independent,

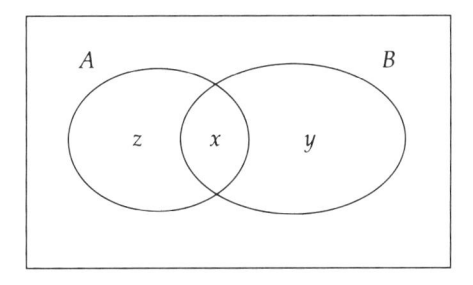

$$P(A \mid B) = P(A), \text{ ie } \frac{x}{x+y} = x + z,$$

and $$P(B \mid A) = P(B), \text{ ie } \frac{x}{x+z} = x + y.$$

The algebra expressions show the equivalence of the two probability statements.

173

Solution

(a)

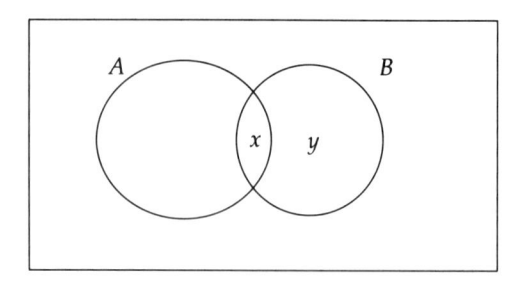

Since A and B are independent, $P(A \mid B) = P(A)$,

ie $\dfrac{x}{x+y} = 0.2,$

$5x = x + y$

$y = 4x.$

But $P(A \cup B) = 0.5$

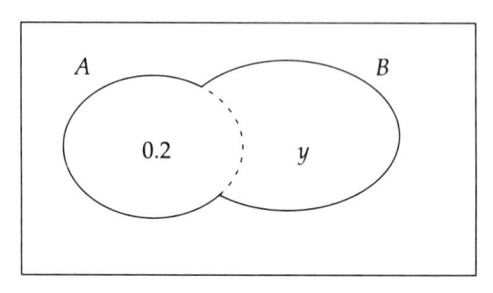

so $y = 0.3$, and $x = 0.075$.

$$\underline{\underline{P(B)}} = x + y = \underline{\underline{0.375}}.$$

(b)

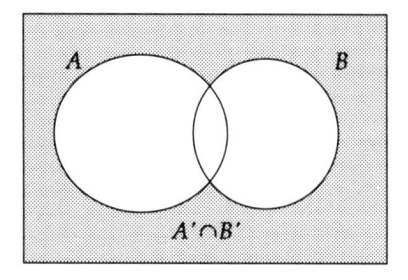

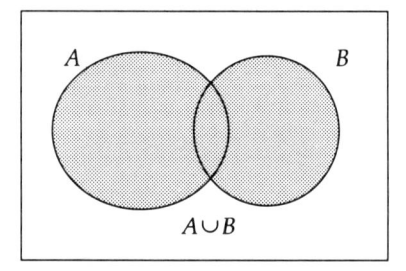

PROBABILITY

From the Venn diagrams, $P(A' \cap B') = 1 - P(A \cup B)$,
so $\qquad P(A' \cap B') = 1 - 0.5$
$\qquad\qquad\qquad = \underline{\underline{0.5}}.$

(c) Since $\qquad P(A \mid A \cup B) = \dfrac{P(A)}{P(A \cup B)}$

$\qquad\qquad P(A \mid A \cup B) = \dfrac{0.2}{0.5}$

$\qquad\qquad\qquad\qquad = \underline{\underline{0.4}}.$

21.3 Dependent events. Given P(A), P(B), P(B | A), find P(B' | A), etc

Question

A and B are two events such that $P(A) = \frac{1}{2}$, $P(B) = \frac{1}{3}$ and $P(B \mid A) = \frac{2}{3}$.
Find

(a) $P(B' \mid A)$,
(b) $P(A \mid B')$.

(L)

Hint Draw a Venn diagram in which the area of the rectangle is 1 unit:

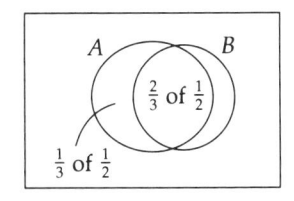

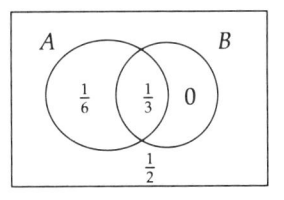

Then the area of region A is $\frac{1}{2}$; since $P(B \mid A) = \frac{2}{3}$, the region common to A and B, ie $A \cap B$, must be $\frac{2}{3}$ of $\frac{1}{2}$, ie $\frac{1}{3}$, and the remaining part of B will be $\frac{1}{3} - \frac{1}{3} = 0$.

Solution

Since either B or B' must happen,
$$P(B \mid A) + P(B' \mid A) = 1$$
ie $\qquad\qquad \frac{2}{3} + P(B' \mid A) = 1,$
$\qquad\qquad\qquad P(B' \mid A) = \underline{\underline{\tfrac{1}{3}}}.$ •

(b) $P(A \cap B') = \frac{1}{6};\quad P(A' \cap B') = \frac{1}{2}$

so $\qquad\qquad P(A \mid B') = \dfrac{\frac{1}{6}}{\frac{1}{6} + \frac{1}{2}} = \underline{\underline{\tfrac{1}{4}}}.$

175

MATHEMATICS REVISION WORKBOOK

Comment

- From the diagram we can see that $P(A \cap B) = \frac{1}{3}$, $P(A \cap B') = \frac{1}{6}$

  so $$P(B' \mid A) = \frac{\frac{1}{6}}{\frac{1}{2}} = \frac{1}{3}.$$

21.4 Use of Venn diagrams

Question

Ann and Bob go out for the evening. The probability that they both go to the disco is 1/12, whereas the probability that neither of them go to the disco is 2/9.

(a) Calculate the probability that Ann or Bob, but not both of them go to the disco.
(b) 'Ann goes to the disco' and 'Bob goes to the disco' are independent events. Ann is more likely to go to the disco than Bob.
  (i) Denoting by α and β the probabilities of Ann and Bob respectively going to the disco, write down the values of $\alpha + \beta$ and $\alpha\beta$ and by solving a quadratic equation, find the probability that Ann goes to the disco.
  (ii) Calculate the probability that Ann goes to the disco, when it is known that just one of Ann and Bob goes to the disco.
(c) On one particular night there were 12 people, including Ann and Bob, on the dance floor. Three different people were chosen at random from the twelve and given a prize. What is the probability that neither Ann nor Bob received a prize?

(AEB)

Hint When we have probabilities for two independent events, it is often helpful to draw a Venn diagram. As the lcm of the two denominators 12 and 9 is 36, let the number of elements in the Universal set be 36, and we can fill in part of the Venn diagram as shown:

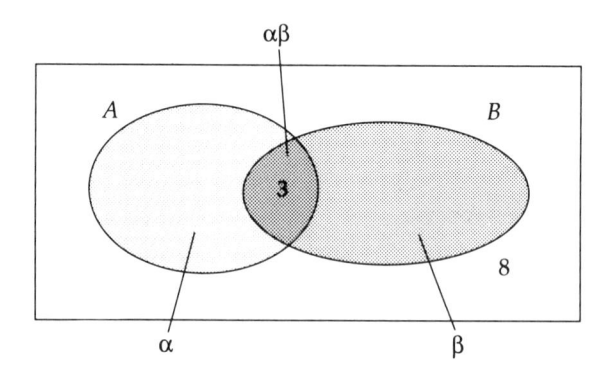

Thinking of α and β as the roots of a quadratic equation, that quadratic will be
$$z^2 - (\alpha + \beta)z + \alpha\beta = 0.$$

PROBABILITY

As Ann is more likely to go to the disco than Bob, the probability that Ann goes to the disco will be the larger of the two roots of this equation.

Solution

(a) From the Venn diagram we can see that the probability that Ann or Bob or both go to the disco is $1 - \frac{2}{9} = \frac{7}{9}$; the probability that they both go to the disco is $\frac{1}{12}$, so the probability that one or other but not both go to the disco is $\frac{7}{9} - \frac{1}{12} = \underline{\underline{\frac{25}{36}}}$. •

(b) (i) We are given that $\alpha\beta = \frac{1}{12}$, and we see from the Venn diagram
that $\qquad \alpha + \beta - \alpha\beta = \frac{7}{9}$
so $\qquad \alpha + \beta = \frac{7}{9} + \frac{1}{12}$
$\qquad\qquad = \frac{31}{36}.$

The quadratic that has roots α and β is $z^2 - (\alpha + \beta)z + \alpha\beta = 0$,

ie $\qquad z^2 - \left(\frac{31}{36}\right)z + \frac{1}{12} = 0,$

ie $\qquad 36z^2 - 31z + 3 = 0.$

Thus $(9z - 1)(4z - 3) = 0$ ••

$\qquad z = \frac{1}{9}$ or $\frac{3}{4}.$

As Ann is more likely to go than Bob, the probability that Ann goes is
$$\underline{\underline{\tfrac{3}{4}}}.$$

(ii) From the second Venn diagram, we can see that now we know α and β we can complete the 'entries' in the Venn diagram, using $\frac{3}{4} = \frac{27}{36}$ and $\frac{1}{9} = \frac{4}{36}$.

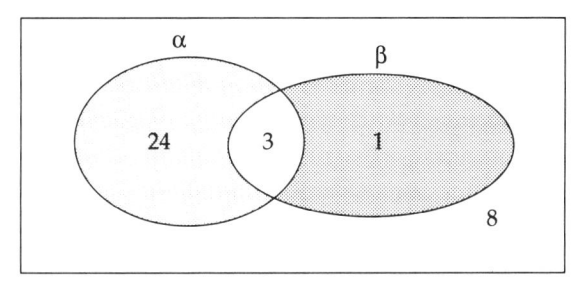

The shaded regions represent Ann or Bob but not both going to the disco, so the probability that Ann goes, given that only one of them is going, is
$$\underline{\underline{\tfrac{24}{25}}}.$$

(c) As Ann and Bob do not receive prizes, the three prizewinners must have been chosen from the other 10 persons at the disco, so the probability that neither Ann nor Bob receives a prize is

ie $\dfrac{\text{the number of ways of choosing three from ten}}{\text{the number of ways of choosing three from twelve,}}$ •••

ie $\dfrac{\dbinom{10}{3}}{\dbinom{12}{3}}$

$\dfrac{\dfrac{10!}{7!\,3!}}{\dfrac{12!}{9!\,3!}}$ ••••

$= \dfrac{6}{11}$

Comments

- We can think of this as
  $P(A \cup B) - P(A \cap B) = \frac{7}{9} - \frac{1}{12} = \frac{25}{36}$.
- •• We could of course use the formula to solve this quadratic equation.
- ••• We may be more familiar with the notation
  $$\dfrac{^{10}C_3}{^{12}C_3}.$$
- •••• If our calculators are suitable we can use them to show $\dfrac{\dbinom{10}{3}}{\dbinom{12}{3}} = 0.545$, to 3 dp.

INDEX

Angle between two vectors   156, 157, 158, 159, 160, 165, 167
Arc length of a circle   58, 59
Area of a sector of a circle   58, 59
Argand diagram   144, 147, 150
Argument of a complex number   144, 145, 146, 147
Arithmetic progression (series)
   common difference, 26, 27, 28
   general term, 26, 27, 28
   sum, 26, 27, 28, 29, 30

Binomial expansion   51, 52, 53, 55, 56, 142

Circle
   cartesian equation, 80, 85, 92, 93, 154
Cobweb diagram   138
Complex roots of a quadratic   18, 150, 152
Concave   19, 61
Cosine
   expansion, 51, 52
   formula, 62, 63
   graph, 20, 77
Cover-up method   43, 45, 47, 48, 50

Dependent events   175
Differential equations   122, 123, 124, 125
Discriminant of a quadratic   19, 22

Exponential
   e^x, 52
   series, 52, 55

Factor theorem   13, 15, 18
Function
   composite, 1, 3, 4, 5, 7, 8
   image under a, 1, 6
   inverse, 1, 3, 4, 5, 6, 8, 9
   one-to-one, 8, 9, 10
   range of a, 1, 3, 4, 5, 6
   sketch, 2, 6, 8, 9, 10, 11, 12

Geometric progression
   common ratio, 26, 28, 30, 31, 32
   sum to infinity, 27, 29, 30, 31, 32
Gradient of a curve   115

Implicit function
   derivative, 100
Independent events   172
Indices   34, 35, 38
Inequality   10, 15, 24
Inflexion, points of   7, 98, 101, 102, 103, 104

Integrals
   by inspection, 109, 111
   by parts, 111, 118, 124
   by substitution, 109, 111
Iteration   135, 136, 137

Line segment (length of)   79, 80
Locating a root of an equation   135, 136
Logarithms
   definition, 34
   derivative, 97, 99, 101, 103
   expansion, 51, 55
   ln, 21, 35, 55
   other bases, 34

Matrix form for representing a vector   160
Maximum value   60, 61, 98, 103
Minimum point   98, 103, 116
Modulus function   10

Newton–Raphson method   135, 136, 140, 142

Parabola   88
Parameters   88, 89, 90, 92, 94
Partial fractions   42, 44, 113
Perpendicular
   distance of a point from a straight line, 167, 169
   straight lines, 80, 81, 162, 169
Plane
   cartesian equation, 156
   vector equation, 156, 165

Quadratic equation   19, 20, 21, 22, 72, 92
Quotient derivative   97, 105

Radians   60, 67, 70, 76
Range of a function   1, 3, 4, 5, 6
Real roots of a quadratic   190, 150, 152
Rectangular hyperbola   88, 89
Reflection in a line   162
Remainder theorem   16
Roots of a quadratic   19, 63

Scalar product of two vectors   155, 161, 165, 168
Secant (graph)   73
Separating the variables   122, 124, 125
Sigma notation ($\sum$)   39
Simpson's rule   130, 132, 133
Simultaneous equations   37
Sine
   expansion, 51, 68, 70
   formula, 62, 64

179

INDEX

Small increments 64, 66
Staircase diagram 140
Stationary points 98
Straight line
   cartesian equation of, 156, 157, 166
   vector equation of, 156, 160, 166, 170
Summation of a series 39, 40, 42
Sums and products formulae 69, 75, 96
Surds 36
Symmetry 104

Tangents
   equation of, 104
   gradient of, 85, 86
Tangent-field 123, 126, 127
Trapezium rule 130, 131

Vector equation of a straight line 156, 158, 159
Venn diagram 172, 173, 174, 175, 176
Volume of a solid of revolution 108, 119, 120